JOHN DEERE
TRACTORS AND EQUIPMENT

VOLUME TWO
1960—1990

Don Macmillan
and
Roy Harrington

American Society of Agricultural Engineers
2950 Niles Road, St. Joseph, Michigan 49085-9659 USA

About the
American Society of Agricultural Engineers

ASAE is a technical and professional organization of members committed to improving agriculture through engineering. Many of our 11,000 members in the United States, Canada and more than 100 other countries are engineering professionals actively involved in designing the farm equipment that continues to help the world's farmers to feed increasing numbers of people. We're proud of the triumphs of agriculture and the equipment industry. ASAE is dedicated to preserving the record of this progress for others. This book joins the first volume of *JOHN DEERE TRACTORS and EQUIPMENT, The Agricultural Tractor 1855-1950, Farm Tractors 1950-1975,* and many other popular ASAE titles in recording the exciting developments in agricultural equipment history.

Library of Congress Card Number (LCCN) 88-71413
International Standard Book Number (ISBN) 0-929355-19-9

Acknowledgements

The contributions of numerous people were used in the development of *JOHN DEERE TRACTORS and EQUIP-MENT Volume Two 1960-1990* during the past two years. Fortunately, many of the folks who were responsible for designing, manufacturing and marketing of John Deere products during the 1960-90 period were available to provide information and to review first drafts. Many retirees see the equipment they helped introduce in the '60s still in use. It was the first opportunity for several engineers to tell the public how this equipment was developed. They described important projects worked on in the '60s '70s and '80s, resulting in many successful products and in some developments that were never offered to the customer.

We are thankful to the following who provided or reviewed material on farm tractors: Paul Browning, Wilbur Davis, Edward Fletcher, Daniel Gleeson, James Kress, Ron Leonard, Ray Meyer, Merle Miller, Sydney Olsen, Bernard Poore, Vernon Rugen, Barrie Smith, Wendell Van Syoc, Ralph Weber and Edward Wright. Help on the text for combines was received from Darwin Bichel, Douglas Bosworth, Kent Cornish, Jerry Hansen, Wayne Slavens, John Wilson and Michael Wyffels. We appreciate assistance from the following on implements: Marv Bigbee, Edward Epperson, Arthur Hubbard, Willard Jenkins, Harold Luth, Glenn Olson, Walter Roll, Bertil Sandin, Fred Stickler and Russell Sutherland.

Consumer products required greater scrutiny because both authors had worked primarily with farm equipment. Martin Berk, Ray Davis and Wayne McClellan thoroughly reviewed and revised consumer products history. Additional information was gathered from Gerald Buelow, Pete Classen and Myron McCunn.

Much of the information on industrial equipment history came from a Deere & Company internal report, *A History of the John Deere INDUSTRIAL EQUIPMENT DIVISION*, written by Brian Alm. Francis Brinkmeyer, Jerald Deutsch, Paul Hanser, Paul Meeden, John Phoenix and Dale Walline provided additional information on industrial equipment. Stanford Barker wrote the first draft of the industrial equipment product review.

We express our gratitude to several individuals who provided information on overseas products: Karl Gorsler, Don Huber and Homer Witzel; to Mabel Buttice and Alberto Souto of Argentina; Bernie Allport, Trevor Rowbottam and Philip Wyndham of Australia; Norman Dunn of Scotland; Mark Moore of Ireland; Gus Gutslaff and Klaus Pajurek of Germany; Luis Sada of Mexico; Bruce Pocock of South Africa; and Angel Barrios of Spain.

Other employees furnished crucial information on multiple product lines or on the company: Brad Boeckner, Robert Bolt, Barbara Cline, Raymond Hollister, Murray Madsen, Wilson McCallister, and William Murray.

Several individuals played key roles in finding, selecting, securing and processing the more than 850 photos in this book. Our prime source was the Deere & Company archives. Leslie Stegh, archivist, runs a well-organized operation that was the major source of advertising and other reference materials. He and Vicki Eller helped find the photos and then process the multitude of orders. Dick Balzer helped in choosing one-third of the photos. His keen eye, interest and product knowledge resulted in good coverage from the photos selected. Many of the photos were also selected from the advertising department photo lab. Cooperation from Delores Hoskins, Jane Huitt, Rick Saltsman and Renate Zerngast of the photo lab helped the book be printed on schedule.

As each chapter of the book was written, editorial scrutiny was provided by Len Lindstrom, Jack Fritts, and by Ralph Hughes, director of Deere & Company advertising. They did the expected job of dotting the i's, crossing the t's and providing a consistent style of writing. More important was their personal interest, understanding of farming, and knowledge of John Deere products.

The team at ASAE, the publisher, deserves a warm note of appreciation for sandwiching work on this book in with their normal work load. Donna Hull, director of marketing, masterminded the operation with able assistance from Suzanne Howard, who followed the details. Bill Thompson displayed his artistic talent in the layout of both text and photos. He also put in the most overtime to keep the book moving. Patricia Howard did the difficult tables. Dee Gunn and Steve Miller gave the book their close examination for errors.

Computers have taken much of the drudgery out of producing a book but they have not eliminated the need for teamwork. We wish to thank the entire John Deere and ASAE team.

Don Macmillan
Roy Harrington

Preface

In the early 1950s, Deere & Company had reached the proverbial "fork in the road." A choice of direction in future product development had to be made. Since the end of World War II, the company had concentrated its efforts on producing tractors, combines and other farm equipment in response to shortages created by the lack of production during the war years. Deere engineers and management began to realize during this period that products designed prior to World War II and manufactured since it ended would not meet the needs of farmers in the decades to follow. Although still extremely popular with farmers across North America, the John Deere 2-cylinder tractors needed to be replaced by a totally new line of tractors that would offer farmers more power, a more versatile hydraulic system and 3-point hitch, and improved operator comfort, convenience and safety... a very tall order indeed.

The decision to abandon the 2-cylinder tractor design was officially made in 1953, but it took seven years of engineering work and field testing, along with many changes in the Waterloo and Dubuque factories, to bring forth the "New Generation of Power" tractors. On Monday, August 29, 1960, Deere gathered all its dealers, and key employees in Dallas, Texas, for the introduction of the new tractor line. This event, known as "Deere Day in Dallas," can accurately be described as the starting point of the Deere rise to industry leader. By the mid 1960s, Deere was recognized as the world's largest producer of farm equipment.

Following the popular and unprecedented acceptance of these new multiple-cylinder tractors came the need to develop a full line of new attachments and farm equipment to utilize the design features of these new tractors. In most cases, bigger was better. Four-row planters, and cultivators were replaced by 6- and 8-row models, and eventually by 12-row behemoths. Harvesting equipment also had to be increased in size and capacity to keep pace with the farmer's demand for increased productivity and efficiency. The 1950s, '60s and '70s saw a rapid transition from 2-row corn pickers to 8-row corn combines. Hay balers that produced 60- to 80-pound rectangular bales gave way to a new "breed" of balers that rolled out one-ton round bales. And the small-grain heads on self-propelled combines grew from the commonplace 12-foot widths to headers that cut 30-foot swaths today.

Volume One of this two-book series covered the development of the John Deere product line from its beginning in 1837 to the end of the 2-cylinder tractor era in 1959. In this second volume, the authors describe how the John Deere farm equipment line was expanded and improved to keep pace with the many changes in worldwide agriculture during the past three decades.

They outline how two totally new product lines were developed and successfully launched: the industrial and consumer products lines. The authors also discuss the John Deere transition from a North American company to a global manufacturer and marketer.

Robert A. Hanson,
former chairman and CEO,
Deere & Company

Contents

PART I–COMPANY HISTORY

PART I

COMPANY HISTORY

1960–1963 The New Generation

A most exciting and far-reaching development came in 1960—the introduction of John Deere's first tractors with more than two cylinders. This vastly improved New Generation of Power was a milestone in tractor history, like Henry Ford's mass-produced Fordson, the Farmall tractor for row crops, and Harry Ferguson's 3-point hitch.

Two-Cylinder Finale

As already indicated in the first volume of this work, a site of over 500 acres at Marshalltown, Iowa, was chosen in 1959 to show the final line of 2-cylinder tractors and the full multiplicity of machines offered to the American farmer.

But one further item was on the agenda. In the large tent sat a fully integral 8-bottom plow. To the amazement of all present, a giant tractor backed in, hooked up, and lifted the plow. It was the 215-hp 6-cylinder 8010 tractor, the first articulated 4-wheel-drive tractor to be offered by a full-line company.

The excitement created had to be experienced to be appreciated. But because the 8010 was so far in advance of anything else available from John Deere, it gave no indication of the following year's even more astounding news.

Deere Day in Dallas

"For a single day, 30 August 1960, John Deere abandoned its conservative formality and played the friendly, expansive host to its dealers and special guests." Thus did George H. Seferovich describe the happenings of that day in *Implement &*

Thousands of John Deere dealers had been flown to Dallas, Texas, to witness scenes which the blacksmith who started it all could not have imagined in his wildest dreams—the unveiling of the New Generation tractors, the 1010, 2010, 3010 and 4010.

Tractor magazine. In settings of pageantry, with closed-circuit television and a $2 million display of equipment, a day dawned at the end of August 1960 that seven years of hard work had been patiently invested in. During that effort, 1956 had seen the opening of the first research and engineering center devoted exclusively to design and testing of tractors by a full-line company.

From the start the 3010 and 4010 were a great success, though the dealers and many of their farmer customers were shaken by the demise of their favorite 2-cylinder "Johnny Poppers." Progress had demanded the change and there was no going back. It was so much easier to mount corn pickers, cotton pickers and similar equipment without fighting the flywheel or pulley of the 2-cylinder tractors.

The giant John Deere 8010 tractor demonstrates its flexibility.

The 4010 diesel tricycle tractor, star of Deere Day in Dallas, soon became a workhorse on North American farms.

Farmer Letters

Farmer response to the announcement of John Deere tractors with more than two cylinders was swift and vocal:

"Why did you abandon the simple 2-cylinder design?"

"Even though your new 4010 may have more power, I need the lugging ability of my 'G' tractor."

"Are the new tractors as good on fuel economy as my 720 diesel?"

"Why did John Deere give up the dependability of its two-cylinder tractors?"

The above comments are typical of those received in the many letters that customers wrote John Deere after the New Generation tractors were announced. They declared legitimate concerns over this major change in tractor design, but also showed an undercurrent of nostalgia for the 2-cylinder tradition. This was best expressed in a letter from one farmer: "Losing 'Poppin' Johnny' is like losing an old tried and trusted friend of the family. It's like losing an old-timer whose advice, a product of his wisdom of many years, we have always found sound."

The tone of farmer comments took a rather abrupt turn as farmers started using the new tractors:

"I like the new 6-cylinder engine. It delivers much smoother power than the 2-cylinder."

"We're well pleased with the fuel economy and, frankly, we have yet to find out how powerful this new tractor really is."

"This tractor is brand-new, but from what I've seen, the variable-speed engine with traditional John Deere lugging power is going to be tops in fuel economy."

"This new power steering is the best I've experienced. It's faster, easier turning than previous systems. This is the most maneuverable tractor I've ever driven."

"How did you ever come up with a seat like this? I never dreamed riding a tractor could be so comfortable. The operator's platform is roomy and controls are easy to reach."

"This is the seventh John Deere tractor I've had and it's the best one yet. It's a nice tractor to operate; handles and rides as comfortably as my car. I'm 78 years old, so this makes a difference."

3

Why Deere Abandoned the Simple 2-Cylinder Design

The handwriting was on the wall of the research engineering department at Waterloo for the demise of the 2-cylinder engine as early as January 1950.

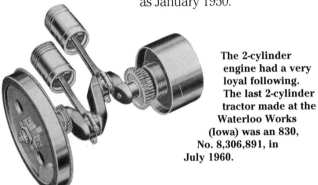

The 2-cylinder engine had a very loyal following. The last 2-cylinder tractor made at the Waterloo Works (Iowa) was an 830, No. 8,306,891, in July 1960.

The first mockup of the "OX" tractor (later to become the 4010) looked like a cross between the Waterloo 60 and the Dubuque (Iowa) 40 tractors.

Many John Deere engineers realized that they had gone about as far as they could go with the 2-cylinder engine design. The cheery pop-pop, miss-miss of the slow-speed engine was music to the ears of many owners. But to engineers it meant poor power balance, which required a heavier drivetrain for equal durability. It also meant that additional power could not be easily obtained by increasing

The 4-cylinder mockup of 1950 continued the Waterloo frame, lights, battery box under the cushioned seat, left brake operated by the left foot, and the traditional belt pulley on the right.

engine speed. Competitive tractor engines were operating at 50-100% higher rpm and getting more power with less displacement. The Model "D" ended its proud 30-year reign as John Deere's muscle tractor in 1953 at 42 horsepower. Its 900-rpm engine's 501-cubic-inch displacement was greater than that of today's 466-cubic-inch engine, which provides the 4955 tractor with 200 horsepower at 2200 rpm. Even the seemingly immortal Volkswagen Beetle design did not last indefinitely!

The '40s in North America were a time of rapid change on the farm. Labor was scarce and high priced during World War II and the following years. Farm income was good, as demand exceeded supply in the ever-expanding export market. With the farmer having adequate buying power, dramatic growth occurred in the use of both fertilizer and farm tractors. Tractors on U.S. farms grew from one million in 1932 to two million in 1943, three million in 1949 and four million in 1953.

The marketplace showed little evidence of a need for more power; the average horsepower of tractors sold in 1951 was 29, little more than the 27 hp of 1943. The John Deere Models "A," "G," "D" and "R" all had more than 29 horsepower, so why were different engines needed? For the past two decades, each additional tractor on the farm had resulted in an average decline of four in the horse and mule population. This switch in farm power was now essentially complete. The replacement buyer was the future tractor market, and he would no longer be attaching horse-drawn implements to his tractor. With the purchase of his next tractor, he would again want a distinct increase in productivity.

How much increase in power could the repeat buyer use? Research engineering set out to determine this by installing a large General Motors diesel in a Case LA standard-tread tractor. At a product engineers' meeting in the early '50s, Wayne Worthington reported that the Great Plains wheat farmer could effectively use considerably more power than anyone currently offered. And he stated that the practical upper limit for 2-wheel-drive tractors was 100 horsepower. However, as hindsight goes, it is not clear if Worthington really felt this was the absolute traction limit, or his credibility limit with management.

In any event, trends in number of tractors sold and average horsepower changed dramatically. Annual tractor sales in the U.S. peaked in 1951 at 442,000 at an average of 29 hp. By 1954 sales were down 54% and power was up 33%.

Manufacturing at Waterloo was as anxious for change as the product engineers. Tooling for the 2-cylinder tractors was wearing out, and manufacturing processes had not kept pace with the changes in technology available after the war.

However, it was probably marketing that convinced management to invest in a complete new line of tractors. The Farmall Works of International Harvester had produced an average of 356 tractors per day in 1950. John Deere implements were gaining market share against IH but tractors were not; they remained below one-fourth of total industry sales. A new tractor line would permit Deere to leapfrog competition in engines, transmissions, hydraulics, and operator comfort.

While the 1950 tractor mockup was gathering dust, a major meeting on new tractors was held in Moline on April 20, 1953. This was followed in May at Waterloo by joint meetings of their product engineering and research engineering personnel.

"Expecting" for Seven Years Without Letting It Show

Bringing an entirely new line of tractors to life without the public knowing it required superb planning, discipline, and a bit of luck. Leaks became more and more likely as an ever-expanding number of employees and suppliers learned of the new program. However, good continuing sales of current tractors demanded that customers, dealers and competitors be kept in the dark.

The decision to start with a clean sheet of drawing paper to develop a new line of farm tractors came in early 1953. To provide privacy, an abandoned grocery store on Falls Avenue in

The May 1954 mockup had the future features of front fuel tank, radiator grilles on each side, planetary final drives, inclined steering wheel, both brake pedals on the right, and deluxe seat on inclined rails.

Waterloo was rented and four trusted employees moved in on April 1, 1953. Few people knew the unmarked building was used by John Deere.

However, John Deere engineers were not hermits, so they soon found their way to the closest doughnut shop. As the number of employees grew, the number of doughnuts picked up each day increased. Finally, the doughnut shop operator volunteered to deliver the doughnuts each morning in time for coffee break. Along with the growth in activity at Falls Avenue came a need for more outside John Deere employees to visit there. These employees were adequately scrutinized for the proper pass at the front desk before passage was permitted. One such visitor was the head of tractor product engineering. As he was being delayed for his pass, in came the doughnut shop operator with a cheery greeting. He walked past the drawing boards and deposited the doughnuts by the steaming coffee pot. A hastily called meeting of the engineers resulted in the decision that in the future, they would carry the daily doughnuts from the front desk to the coffee area.

Drawings made at Falls Avenue did not include the usual John Deere Waterloo Tractor Works identification. Drawings were sent to many job shops over a wide geographic area. A limited number of parts went to each shop, so

Multiple test tracks, including a mud bath, were located on the farm.

they could not "put two and two together" and detect what John Deere was doing. The codes for the tractors to come were "OX" for the 4010, "OY" for the 3010, and "OZ" for the 5010. The first "OX" was assembled in the spring of 1955. The Tractor Works had used land next to the sheep feeding lots of Rath near Waterloo for experimental field testing. Although this was adequate for local testing during the gradual evolution of the 2-cylinder line, it did not offer the security required for the new line. Therefore, a 684-acre farm a few miles southwest of Waterloo was purchased. The first shop building was completed in 1954, and the engineering office was occupied in September 1956.

Complete new designs require many hours of testing to ensure both function and durability. The new Product Engineering Center (PEC) farm had a limited amount of farmland out of public view, and farming could not be done in the winter. Therefore, in the spring of 1956 two experimental tractors were disassembled and boxed for shipment to a test site in southern Texas. One engineer, Wendell Van Syoc, and one mechanic, Gene Lonergan, were chosen with care to make the trip and reassemble the tractors. To help keep their mission secret, they used an old unmarked Reo truck.

The first day of the trip was uneventful. However, as the second day wore on, so did the truck's engine. Finally, it was missing so badly and losing so much power that it had to be

repaired in Waco, Texas. The truck mechanic found a burned exhaust valve, and no replacement was available. Fortunately the mechanic was clever enough to recognize the similarity between that valve and a John Deere Model "B" tractor valve. Shortening the "B" valve required additional time, so the truck had to be left in the shop overnight. Van Syoc was quite concerned over maintaining security with the truck left in a public garage. But his evasive answers warded off questions about the cargo and its destination. Had the trip been made more recently, the secret contents of the truck probably would have been searched for drugs.

Other tractor projects at Waterloo helped obscure some of this development program. The 2-cylinder 30 series tractors that came out in August 1958 had distinctly softer curves than their straight-line, square-appearance predecessors. This much difference in appearance

Test tractors were camouflaged, with red bodies and yellow wheels. Side shields were used to conceal the distinctive silhouette when run on the test track.

indicated to dealers and customers that no additional major change was to be expected in the near future. The many strange parts and activities related to the 1959 introduction of the 4-wheel-drive 8010 tractor kept employees from recognizing that some experimental parts were for an entirely new line of tractors.

To provide a continuous supply of 2-cylinder tractors to the dealers, an extra large volume was made and stored in an alfalfa field in May 1960, prior to shutdown for retooling. When the new line was finally introduced to the dealers at Deere Day in Dallas in August 1960, it came as a complete surprise. While the dealers were there, trucks and railway cars started rolling out of Waterloo, with the new tractors covered by large cardboard cartons so they would arrive at the dealerships undetected by the customers. The great effort to keep the new tractors quiet had succeeded!

V Engines

When John Deere decided to abandon the 2-cylinder design, its only experience at building engines was with 2-cylinder tractor engines and single-cylinder, hopper-cooled stationary engines.

The Dubuque Works started building 4-cylinder gasoline engines in 1953 for the 114W and 116W hay balers and the 25 combine. Later 6-cylinder units were added for the 55 and 95 combines. Although this integral-bore type of gasoline engine was adequate for balers and combines, the basic design was unsatisfactory as a diesel engine for tractors.

This led to a major redesign to include wet sleeves for longer life in the 1010 and 2010 tractors. A novel construction offered many potential advantages. Instead of four individual wet sleeves with their inherent sealing problems, a sleeve-and-deck insert combined all the sleeves into one unit. But the promise of this clever design was never quite fulfilled, and the engine was made for only a few years. The diesel version did not have the desired durability for tractor usage.

A year after the 1010 and 2010 were introduced, Waterloo engineers began designing a diesel engine to be used in tractors manufactured in Mannheim for the European market. As the design progressed, it was decided to use this basic engine worldwide in the 1020 and 2020 tractors. A gasoline version was added for sales in the U.S. and Canada. The 300 family of 3-, 4- and 6-cylinder engines remains in use today in the 2155 through 3155 tractors.

This Dubuque Works design simplified both the casting of the block and assembly of the engine.

Probably the only carryover for Waterloo from the starting engine background to tractor engines was the desire to break with tradition and join the trend of the times. Even Chevrolet, the conservative market leader in cars, started offering a V-8 engine in 1955 as an option to the straight 6-cylinder engine that had been their standby since 1929. V engines offered two distinct advantages for John Deere's new line

The New Generation

The high-speed V-4 starting engine, introduced on the 70 diesel in 1954 and the 80 diesel in 1955, gave Waterloo its first experience at designing and building engines of more than two cylinders.

This January 1956 "OX" tractor is the first of five second-build units.

of farm tractors. The engines could be shorter, permitting higher-power tractors without excessive length. Manufacturing and service would be simplified if V-4, V-6 and V-8 engines shared common parts. A gas V-6 engine was shown in the January 1955 "OX" mockup.

In 1956, the exhaust manifolds had been relocated to the outside of the engine. But the heat on the operator's feet and legs was intolerable, since the front-mounted fuel tank located the engine immediately in front of the operator. So heat shields were located over the manifolds to reduce the heat from the fan blast and radiation.

The 2-cylinder tractors with their tapered fuel tanks had provided an excellent view of front-mounted cultivators. The new V engines needed to be narrow to offer somewhat comparable visibility and also to have reasonably good

appearance. To meet this requirement and fit within a narrow frame, a non-conventional 45-degree V was chosen.

But it had drawbacks. This design required six separate throws of the crankshaft on the 6-cylinder engine for even firing. Adequate crankshaft bearing life could not be obtained in the diesel version. Other problems with the V design included cost, heat on the operator, and reduced visibility for cultivation. Dropping the V design meant that a new larger 6-cylinder engine had to be designed in place of the V-8 planned for the "OZ" or 5010 tractor.

The July 1956 "OX" mockup showed an in-line 6-cylinder engine. This engine was still required to fit into the 16 5/8" narrow frame, and thus had a compromise on the location of the camshaft. A novel feature of this engine and its V predecessor was crankshaft rotation in the opposite direction from traditional engines, to save one gear in the power train.

The tractor frame was widened to its current 20" in the 1958 build of experimental tractors. This permitted the redesign of the engine into what is known today as the 400 series.

Because of the quality of 300 and 400 engine designs, for a number of years John Deere has been the largest North American producer of diesel engines for off-highway use.

The first "OX" tractor was completed in April 1955 with a V-6 diesel.

Henry Dreyfuss considered each visible component and good operator visibility in the styling of the 3010, 4010 and 5010. It proved distinctive and set the pattern for the next decade. The two European tractors were similarly styled.

The New Generation Plans Develop

The criteria for these new machines were laid down in advance: to be commercially successful the new tractors must satisfy four groups of people, the first and most important being the farmer customers. They needed a design that would give them more speed, more efficiency, and more precise control of power farming. And the operator's comfort was a major consideration in achieving those ends.

The second group to consider were the manufacturers of implements and equipment to be used with the new line. The setting of new standards in hitches, hydraulics and PTO drives was necessary here.

Then the company's marketing division must have the various options it required, and a striking appearance to help sell the line. A family resemblance would be a great advantage.

Also, the factories that would build the new tractors would welcome a minimum number of models, with the greatest possible inter-

changeability of parts. If their existing skills could cover the new line's building requirements, this too would be useful. Consultation with these four groups throughout the seven years proved to be the reason for the new line's instant success.

The new tractors were truly to be built for the future. The multicylinder variable-speed engines meant they could be operated at a speed to suit the job, rather than at full throttle as was the case traditionally. Machines needing a high power output from the PTO were best operated by the 1000-rpm outlet, while new mid-mounted machines could use the new front PTO of the two larger tractors and those built in Mannheim.

The new transmission shifted collars rather than the gears, which remained in constant mesh. The common reservoir of oil for the gearbox, rear axle, and all the hydraulic operations was a first on farm tractors.

The hydraulic system had a much increased capacity to cope with these many calls simultaneously. The exclusive Load-and-Depth Control system for the 3-point hitch allowed automatic load control, depth control, or both.

Model Sizes Chosen

The New Generation of Power was to begin with two sizes from Dubuque and two from Waterloo. The smallest model was the 35-hp 1010 with gas or diesel engine. It came in single-row-crop, utility, row-crop utility, or crawler mode for farming, and in wheel or crawler industrial types. The 45-hp 2010 was available with gas, diesel or LP-gas engine options as a row-crop with dual-wheel, Roll-O-Matic, single-wheel or wide-axle front end choice. It also came as a row-crop utility, a Hi-Crop or a crawler for farming, and in industrial models similar to the 1010. These two series were designed and built at Dubuque. They retained the fuel tank above the engine, had dry-disk brakes, and offered power steering as an option.

The 1010 was little more than a 430 with a new engine, plus sheet metal for family styling. Thus, customers found this proven design very dependable.

The 2010 was an entirely new tractor with many of the features of its Waterloo cousins. Unfortunately, it did not receive the same amount of engineering time and attention and thus was destined for some durability problems.

The orchard version of the 3010 had a lowered operator's station to provide a sleek profile.

The 4-cylinder, 55-hp 3010 shared many common parts with the 6-cylinder 80-hp 4010. Both offered diesel, gasoline or LP-gas engines. Both were available as standard models or as row-crops with a choice of Roll-O-Matic, dual-wheel, single-wheel, or adjustable-axle front end. In addition, the 3010 was offered as a row-crop utility and the 4010 as a Hi-Crop. Both sizes were also available as industrial units. The industrial tractors largely followed the styling established in 1958 with the previous models, except that the radiator grilles were steel plates instead of castings.

In 1962, the 1010 and 3010 series were extended by the grove and orchard models. The 5010 was also announced, the highest power 2-wheel-drive tractor on the market. Some U.S. personnel were amazed when the United Kingdom importers requested this model for announcement at the 1962 Smithfield Show in addition to the 4010.

The 5010 was the first 2-wheel-drive tractor with over 100 drawbar hp, enough to pull the 7-bottom F245H plow.

THE Furrow

SPECIAL EDITION / SEPTEMBER · OCTOBER · 1960

ntroducing a new generation of power

William A. Hewitt, president, Harley Waldon, Waterloo Tractor Works manager, Maurice Fraher, vice president, and Merlin Hansen, chief engineer, admire the new tractor line.

All the Comforts of Home

The "clean sheet of drawing paper" approach to the new line of tractors permitted engineers working on operator comfort and convenience to change the image of John Deere from being behind competition to being the distinct leader in the field. Several things were less than ideal on the 2-cylinder Waterloo row-crop tractors. The transmission was difficult to shift, requiring a crosswise motion to slide large gears. The brakes required good coordination of both feet to achieve smooth stops from high speeds. The hand clutch was not like the clutches of cars and trucks.

Partially for security reasons, the product development department in Moline was chosen in 1953 for the early work on the operator's station. The Henry Dreyfuss group of industrial designers had worked with John Deere since 1937 but this was their first real opportunity to do much for the operator. They obtained the services of Janet G. Travell, noted orthopedic doctor, to assist Deere in the development of the seat and suspension.

The relationship between the steering wheel, pedals and seat was determined for short, average and tall operators on a stationary mockup. Car seats had traditionally been adjustable, to the rear and down or at least horizontally, so the tall operator had enough headroom. Deere resolved to provide a good fit between the seat and controls for all operators from the short farm boy through the tall farmer. It was obvious that inclining the seat track up and to the rear would provide better pedal operation for the short person and a more comfortable knee angle for the tall person.

Standing provides some relief after sitting for long periods. The seat was designed for this. It moved to the rear when a latch was released, and then returned to the preselected height adjustment when the operator sat down. The seat cushion was designed for comfort, with enough foam to reduce engine vibrations. The lower backrest provided secure side support for the operator.

A definite step forward was also made in the design of the seat suspension, providing a good ride during normal rough field work and road

11

transport. The spring and shock were adjustable together to give an optimum ride for a wide range of operator weights.

Gauges were grouped for easy reference and controls were placed within easy reach. Hydraulic and PTO controls were located for operation by the left hand, throttle and gearshift by the right.

Access to the roomy, unobstructed platform was moved from the rear of the seat to the front. Convenient steps and handholds were located on the left and right. Optional large dual-headlamp fenders gave the operator better protection from dust, mud, and contact with the tires. There was a good view of front-mounted cultivators, row markers, rear-mounted equipment, draft links and drawbar.

Marketing was not certain that the customer was willing to pay for all this comfort and con-

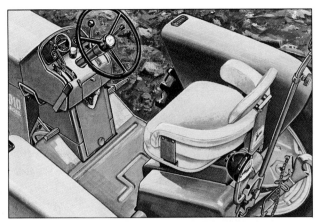

The 3010 and 4010 set new standards for operator comfort and convenience.

venience, so they offered the deluxe posture seat and suspension as a $50 option. Within a few years 95% of customers were choosing that option, so it became standard equipment. During the next two decades, variations of this seat and suspension were used on more than 90 models of John Deere equipment.

Ride comfort took another distinct step forward in 1977 with the introduction of the 40 series tractors. The Personal-Posture seat was provided on those tractors with the now widely accepted Sound-Gard body. The seat offered an adjustable backrest and armrests along with a cloth surface for better breathing and a more secure feeling. Separate levers were provided for vertical and horizontal adjustments for operator

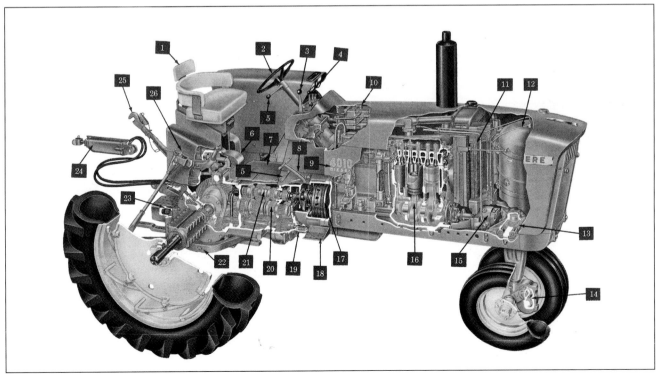

Deere invested $50 million in the development and tooling for the New Generation of Power. Many of the features shown here remain in use in today's tractors.

height and weight. Vertical vibrations were reduced with a hydraulic-over-air suspension and adjustable damping. An innovative attenuator reduced fore-and-aft vibrations in rough field work.

In 1990, almost all Waterloo tractors were sold with Sound-Gard body. However, those few that were sold for minimum price to owners with hired operators were equipped with 4-post Roll-Gard structure. On these tractors, the basic seat and suspension that was introduced 30 years ago was still used.

Hydraulics Replace Muscle

Much of the operator comfort and convenience was provided by the unique hydraulic system of the 3010 and 4010 tractors. John Deere had been a pioneer in having the tractor, rather than the man, supply the power to lift implements. The first mechanical power lift was introduced on the 1928 Model "GP" tractors. This was followed by the 1934 Model "A," which offered a hydraulic rockshaft to lift cultivators and other integral equipment. Although a small manufacturer had introduced remote hydraulic cylinders earlier as an add-on kit, Deere was the first tractor manufacturer to engineer such a system. The Powr-Trol system was introduced for drawn implements in 1945.

Mounted corn pickers and cultivators made tractors very difficult to turn at the end of rows. In 1954, the John Deere 50, 60, and 70 tractors were the first to offer integral power steering.

It is difficult to say what part of the total new tractor design represented the most dramatic and long-lasting influence on future design, but the hydraulic system was certainly near the top. Engineers wanted more hydraulic functions but a simple design. A closed-center system was chosen because it offered power on demand at multiple locations with a minimum of power drain from the engine when not being used. The 8-piston variable-displacement pump was driven from the front of the crankshaft. The transmission and differential case served as the reservoir for the single oil system for both hydraulics and gear lubrication. A filter kept the oil clean, and an oil

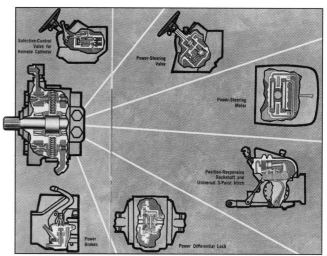

A single pump provided adequate hydraulic power at multiple outlets.

cooler kept the temperature reasonable for proper lubrication.

The power steering was unique in that for the first time on tractors, there was no mechanical connection between the steering wheel and the front axle. If the engine failed on the road, the steering wheel operated a manual pump to provide emergency steering.

The wet-disk power brakes were enclosed in the differential housing, which reduced wear and kept them free of water and dirt. A special oil was developed to provide good gear lubrication and smooth wet-disk brake operation. Individual brake pedals were used for tight turns at row ends, but were designed to provide equal pressure on both brakes for straight stops on the road.

New breakaway couplers were used on the optional one or two remote cylinders. One or both of these cylinders could be used on a front rockshaft to control mounted equipment.

Up until this time, tractor manufacturers had used the top link of the 3-point hitch to sense draft, just as Ferguson had when he introduced it. This depended on the implement wanting to overturn with increased draft loads, and worked reasonably well with small implements. Deere had already encountered problems with this design on the larger 2-cylinder tractors when used with plows and disks. The solution was to use the lower links to sense draft. These links were attached to a frame which caused a bar to bend, sending a signal to the rockshaft control valve.

Old implements could be used on new tractors and new implements could be used on old John Deere tractors.

So successful was the closed-center hydraulic system that its basic design can be seen in the 55 series tractors sold today. But now it also powers the differential lock, Power Shift, Perma-Clutch and the Personal-Posture seat.

Quik-Coupler

"About all a tractor can do without an implement is to go after the milk cows" was a favorite saying of Russ Sutherland, a recent Deere & Company vice president of engineering. Designers of the new tractors had two distinct challenges. The first was to design a tractor that would accept all the current implements the farmer owned for mounting on his 2-cylinder tractors. The second was to make it easier to attach implements, especially as they were getting too heavy to manhandle into position.

With corn as the most valuable crop in the U.S., major consideration was given to attaching front-mounted cultivators and 2-row corn pickers. These machines surround the tractor and require good visibility. The bolt holes in the side frames of the new tractors were made as similar to those in the old tractors as was practical. Movement of the seat forward, for a better ride, was limited by a compromise favoring a clear view of the front-mounted cultivator.

Loader frames often attach to the rear axle to provide strength. However, the loader interferes with other implements, so it needs to go on and off quickly and easily. A design was developed that worked similar to locking pliers, clamping diagonally on opposite corners of the rear axle housing without the use of wrenches. The design worked well in the field but met its downfall due to lack of interchangeability. New implements would not fit on old tractors and old implements had to be converted to fit the new tractors. An examination of 3010 or 4010 axle housings will reveal two pockets near their outer ends. The design was so near adoption that the tooling people had already chosen these clamping pockets to locate the housing for machining.

Engineers had more success in developing an easy method of attaching implements to the 3-point hitch. Plows and other mounted implements had become larger and more difficult to attach. The operator had to get off and on the tractor multiple times to make the connection. The Quik-Coupler attached to the tractor 3-point hitch and simply moved the implement back a few inches. Following this introduction by John Deere, engineers from various manufacturers and other engineers worked together to develop an ASAE standard, in December 1964, on quick-attaching couplers for 3-point hitch implements. This provided for compatibility between new and old as well as between the various brands of tractors and implements.

Deere's First Overseas Company

William A. Hewitt had succeeded Charles Deere Wiman in 1955 as Deere & Company president. As his running mate on the financial side Hewitt had Ellwood F. (Woody) Curtis, and the two were to form a formidable team over the succeeding years, Hewitt with his broad vision of the company's destiny and Curtis with his financial expertise. The following year Curtis went to Mexico. A branch was established and land for a factory was purchased before the end of the year. A new company was formed, John Deere CA, registered in Venezuela, to cover this overseas expansion.

Later in 1958 the U.S. 630, 730 and 830 tractors were assembled in Mexico, and early in 1959 the new factory in Monterrey, Mexico, was opened. Thus the company's overseas arm was beginning to take shape for the momentous events about to hit the agricultural machinery world.

The German Story

To retrace our steps for a moment, developments had taken place elsewhere which would have considerable effect on the growth of the company into the largest agricultural machinery producer in the free world. The need to expand abroad if the company was to keep up with or overtake the opposition was apparent. In 1953, the year in which the tractor research team was organized, Lloyd Kennedy investigated the possibility of purchasing Heinrich Lanz, A.G. in Germany, but the board decided against it.

Lanz first built a tractor, the Laudmaumotor LB, in 1911. It was 1921 when the first famous Bulldog single-cylinder tractor of 12 hp was introduced; two examples of this model are to be found, one with the U.S. archives and one in the Mannheim factory museum. In 1929 came the HR5 15/30 and HR6 22/38, the first radiator-cooled models.

The first Lanz Bulldog tractor featured a single-cylinder, hopper-cooled engine and hard rubber tires.

Just before the war in 1939 Lanz introduced a new row-crop type, the HN3 Allzwerk or Allwork, and in 1942 the 100,000th Bulldog was built. Soon afterward the war reduced the works to rubble except for the water tower—a distinctive landmark in Mannheim.

Lanz introduced a completely new concept during 1951 in the Alldog A 1305. This self-propelled toolbar had a 12-hp gasoline (or benzine as it was called in Germany) engine. The company introduced a range of semi-diesel (start on gasoline, run on diesel) tractors in 1952-53, starting with the three models D1706/1906/2206 and followed by three more, D2806/3206/3606. The D2806 was also available as a twin-front-wheel row-crop tractor. The 150,000th Bulldog was produced on February 9, 1953.

In the same year Lanz, which had experimented with a combine in 1938, announced their first model, the 6' PTO-driven MD 180, followed in 1954 by an 8' self-propelled model, the MD 240 S.

Another Lanz development at this time was the establishment of a factory at Getafe, near Madrid, Spain, in 1953. In its first year of production, 222 Bulldog 36-hp tractors came off the line.

The last single-cylinder Bulldog built in Mannheim was No. 219,253, this being the number of Bulldogs built from 1921 to 1960.

A second look at Lanz was taken by M.A. Fraher, and this time Deere & Company decided to purchase, effective November 12, 1956. At the time, the German Lanz line consisted of 19 models from 11 to 60 hp, but within a year these had been reduced to 13.

All Lanz Bulldogs were painted green and yellow in 1957-58 and carried the name John Deere-LANZ. By 1959 two further models had been dropped and the D3606 had given way to the D4016 40-hp model.

Two new tractors, the 300 and 500, were announced in Mannheim in January 1960. These were diesel 28- and 36-hp models with an entirely new look, a 10-speed Lanz gearbox, sprung front axle and rubber-mounted engine. The novel

John Deere-Lanz 500 tractor No. 104,611 at a combine demonstration in 1962.

sleeve-and-deck 4-cylinder engines were the type designed for the soon-to-be-announced 1010 tractors from Dubuque, de-rated in the case of the 300.

The 300 and 500 tractors had been styled by Dreyfuss & Associates and were the size in demand in the European market. They had been flown to the United States for testing and were put into full production to replace six of the smaller Lanz models. To cover the crawler requirements for Europe the 440 was produced, but with a 3-cylinder Perkins diesel engine.

During 1962, the 2-cylinder (vertical) 100 tractor (18 hp) and the 700 (50 hp) with the 2010 engine were added to the line. Note the German-style fender seat for a passenger.

During 1961 the company increased its holdings in Lanz Iberica (Spain) and eventually acquired control. The next year they had manufacturing facilities in six countries outside North America—Germany, France, Argentina, Mexico, Spain and South Africa.

Lanz in Zweibrücken, Germany, had been building 214 series balers since 1958, as well as 15S combines.

For 1963 two additional tractor models were added to the French line, the 303 and 505 with 37 and 44 hp respectively, and on them the Lanz name was dropped.

Argentina Joins the Story

From the beginning of the century Agar Cross had been Deere & Company's agent in Argentina, and late in the 1920s they had introduced the Model "D" to that market. In 1958 the Argentine

Hay un NUEVO JOHN DEERE 730

Argentina made the 2-cylinder 730 tractor from 1958 to 1970.
Note their different medallion design due to a trademark
conflict.

was used for assembling some
of the U.S. industrial line.

At this time C.C.M. was
building the D25 and D28 combines. For haymaking, they
produced six small low-density
balers.

South Africa Added to the Team

In 1962 John Deere was
manufacturing in South Africa,
having purchased a 75% holding in South African Cultivators
(Pty.) Ltd. This company, which
had a plant at Nigel in the
Transvaal, became known as
John Deere-Bobaas (Pty.) Ltd.,
but this was changed later in
the '60s to John Deere (Pty.)
Ltd. During this period the
product line marketed grew to include tractors,
combines, and forage and hay equipment, mostly
sourced from Europe.

operation was boosted with the opening of a plant
in Rosario to manufacture up to 3,000 tractors a
year. A percentage of their content had to be local,
and the first models built there were 2-cylinder
730s. Rosario also built the 445 tractor from 1963
to 1970. It was based on the Dubuque 435, which
used the 2-cylinder General Motors diesel engine.

French Cooperation

In France, in 1958, the three farm machinery
companies Remy of Senonches, Rousseau of
Orleans and Thiebaud of Arc-les-Gray had joined
together to form Compagnie Continentale de
Motoculture or C.C.M. Deere & Company decided
to join this marketing organization the following
year.

During 1962, Deere also purchased a site at
Saran, near Orleans, France, where an engine
plant for the European factories was to be built.
The production of the Rousseau plant having
ended or been transferred elsewhere, this plant

A Worldwide Tractor Concept

As already mentioned, the Dubuque tractors
and the Mannheim equivalents had the novel
sleeve-and-deck engine with two cylinder-head
gaskets, one below and one above the deck.
These were the only models so fitted, the next
series reverting to the normal arrangement.

With the expanding markets abroad, planning
of a worldwide tractor began in September 1960,
as soon as the new models had been successfully
launched. The consensus view was that two
models between 25 and 50 PTO hp should be
developed, and the following year the experimental 3-cylinder X21 (37 hp) and 4-cylinder
X22 (55 hp) tractors were planned.

But the 1963 season was to see further
advances for Deere, which was that year to
become the world's number one producer of
agricultural equipment.

1963–1972 A Classic Tractor

The 4020 tractor is a classic in every sense of the word. Good 4020 tractors still bring more than their original price at farm sales. There have been more 4020s made than any other John Deere since the 2-cylinder days.

Power Shift for 3020 and 4020 Tractors

In the fall of 1963 the new 3020 and 4020 models were announced and sent to Nebraska for testing, and at the same time became available on the Mexican market. They were immediately successful. The 6-cylinder 4020 must be ranked with Fordson, Farmall and Ferguson as one of the four tractors that have most influenced world tractor design.

Power was increased for the 4-cylinder models from 59 hp at the PTO to 65 hp and on the 6-cylinder from 84 to 91 hp. Another great improvement, an optional Power Differential Lock on all the Waterloo models, came in 1964.

In 1965, to complete the 20 series lineup, the 5010 became the 5020 with increased power from 121 to 133 hp. A further upgrade in 1969 brought the 5020 to 141 hp.

When the worldwide 1020 and 2020 tractors were introduced, they no longer offered the

4020 gasoline tricycle tractor with Roll-Gard ROPS.

An optional Power Shift transmission provided eight forward and four reverse speeds without clutching.

The "Rusty Palace"

The general offices of Deere & Company had been located at 1325 Third Avenue, Moline, Illinois, for 85 years before William A. Hewitt became president in May 1955. The 4-story office was boxed in on all sides by taller John Deere factory buildings, leaving no opportunity for further expansion. As more employees were required to cope with the ever increasing business, additional office space was rented in downtown Moline.

With some confidence that the postwar boom in farm equipment was here to stay, the company had built a new office building a mile east at 3300 River Drive. Advertising, patent and purchasing departments had moved there in August 1954, solving the immediate problem of office space. This 2-story building was then considered the beginning of a new general office, as the several-acre plot would allow considerable expansion and adequate employee parking.

Hewitt, however, had a larger vision for the future of the company. This vision was defined in

tricycle row-crop front option of the Dubuque 2010 or the Waterloo 3020 and 4020. The 2510 filled this gap admirably at 55 PTO hp by the installation of the Dubuque 2020 gasoline or diesel engine in a 3020 chassis. Both Syncro Range and Power Shift transmissions were offered. The 2510 became the 2520 in 1968, with power increased from 55 to 61 hp.

The row-crop farmer's thirst for power was met by offering the 5020 with adjustable front axle and dual rear wheels.

At least three additions to the original 1870 Deere & Company general office were visible in 1950.

Saarinen captured the company's strength and roots in the soil by using an external steel frame. This was the first major building to use a corrosion-limiting steel that requires no painting to maintain its rich earthen tones. This appearance, along with the high-quality offices having scenic views of nature, led some people to coin the term "Rusty Palace." For those who design, build and sell John Deere equipment, the building stands as a symbol for company excellence.

Fill dirt for the product development and research building had been placed a quarter mile south of the main buildings, across the highway. However, construction of this building was postponed until the other two could be completed. When the majority of general office personnel moved to the new $8 million Administrative Center in April 1964, the two departments working directly with equipment design remained behind. The engineering research department (previously product development), which had been at 301 Third Avenue, moved in July 1964 temporarily to the recently vacated building at 3300 River Drive, now the Technical Center. With

a public announcement on August 14, 1957. Deere & Company had purchased some farms 3 miles south and 4 miles east of their downtown Moline office. Eero Saarinen, world renowned architect, had been engaged to design the company's new worldwide headquarters. Three buildings were announced for the new Administrative Center: an office building, an auditorium and display building, and a product development and research building. This was to be the new home for 850 employees from the main office, the 3300 River Drive office, and various rental locations.

The shared vision of Hewitt and Saarinen resulted in a masterpiece of architecture. Hewitt wrote,

"The men who created this company and caused it to grow and flourish were men of strength—rugged, honest, close to the soil. Since the company's early days, quality of product and integrity in relationships with farmers, dealers, suppliers and the public in general have been Deere's guiding factors.

"In thinking of our traditions and our future, and in thinking of the people who will work in or visit our new headquarters, I believe it should be modern in concept, but at the same time, be down to earth and rugged."

Saarinen nestled the 9-story Administrative Center in an oak-lined valley. Many visitors tour the building each year.

Employees have a panoramic view of nature uninhibited by the usual clutter of cities.

a decline in the economy, the temporary home became permanent for engineering research. Materials engineering remained in the top floor of the Plow Works until 1976, when they also moved to 3300 River Drive.

Excellent sales in the '70s resulted in more employees and the need for more space. Kevin Roche-John Dinkeloo & Associates, from Saarinen's original firm, continued the theme of the original building in the expansion to the west. The 3-story West Building featured an atrium and was occupied in June 1978. In 1990, there were about 1400 employees in the two buildings. The rapidly expanding insurance business resulted in another building of entirely different architecture on the same 1,000-acre plot but nearly out of sight of the Administrative Center. It was occupied in 1981 and now accommodates about 400 employees.

The old main office was torn down soon after

it was vacated. In May 1989, many of the surrounding oldest Deere factory buildings north of Third Avenue were demolished. The 20-acre parcel they were on had been given by Deere & Company to the city of Moline as the site for its Civic Center. Some Deere & Company Distribution Service Center employees remain at 1400 Third Avenue, diagonally across from the site of the old main office. Deere Archives also remains at the old John Deere Spreader Works at 1209 13th Avenue, East Moline.

Vision for The Long Green Line

From the company's point of view, 1963 saw the commencement of one of their most successful innovations. The introduction of the 7-hp 110 lawn and garden tractor started today's consumer products division, which has the world's leading line in that sector of the market. This field has grown so large that we will deal with it later in a separate chapter.

It was coincidental that in 1963, during the construction of the company's new headquarters, Deere passed International Harvester to take first place worldwide in total sales of farm and light industrial machines.

April 1964 saw the advancement of William A. Hewitt to chairman and Ellwood F. Curtis to president. This was the well matched team that led the company for its two decades of greatest growth. Hewitt's vision took John Deere to new heights until his retirement in 1982. Curtis, with his wise counsel and vast experience, helped make Hewitt's best dreams come true. His tenure lasted until retirement in 1978.

Two major introductions were witnessed in 1965—a new line of combines designed in and for Europe, and the worldwide-design 1020 and 2020 tractors announced in the U.S. and Canada.

Power Train '66 was the biggest announcement program in John Deere history. A three-quarter-mile-long train roamed the United States to show off "The Long Green Line" of tractors and 43 new implements.

John Deere sales continued to grow in the '60s, hitting $1 billion for the first time in 1966.

During 1967, the first industrial sales branch was opened in Baltimore, Maryland, indicating the rapid increase in this side of the business. Since this line grew to be a story in itself, it will be dealt with later in another chapter.

Combine Designed in Europe

During harvest 1961 a team from Harvester Works led by Homer Witzel had been looking at the European combines operating in the field. Of particular significance was the similarity of some European models to the John Deere 55.

In fact an example of these could be found in both the Claas and Claeys (later Clayson and then New Holland) works. In the latter case their first self-propelled model was effectively a 55 widened in the separator from 30" to 40" to cope with the extra straw grown in Europe. The first Claas self-propelled was called the Model SF.55, a compliment by any reckoning.

A European 630 combine with cab is shown picking up windrowed grain in western Canada.

When the new line of John Deere combines emerged from Zweibrücken in 1965 they were in fact an updated version of the 55 with the extra width required for local conditions. The basic layout had a center platform with grain tank behind, and the engine behind that.

The first models announced were the 4-walker 530 and 630, identical except for their engines, a 4-cylinder 79-hp on the 530 and a 6-cylinder 100-hp on the 630. They were followed the next year by the smaller 3-walker 330 and 430 self-propelleds, the tractor-drawn PTO-driven 360, and the top model of the line, the 5-walker 730 with a 6-cylinder, 115-hp engine.

All the engines were diesel, produced in the newly opened engine factory in Saran, near Orleans, France. Tanker models were standard, with sackers as an option, and the machines were an instant success. Much of this was due to the 24" diameter cylinder with 104-degree concave wrap; the 6" throw on the straw-walker cranks for slower, more vigorous agitation of the long straw; the walkers' length (143" on the 430 up); and the opposed action of chaffer and sieve.

The result was minimal grain loss, compared with that of competitors with 4" walker throw and unison sieve action. In addition the comfort of the operator had been considered and all the controls came easily to hand. Dealers in the U.K. were very pleased to have this addition to what was at that time a short line in the British market.

1020 and 2020 Tractors Introduce Worldwide Design

To add to this progress, late in 1965 the 1010 and 2010 tractors, which had not been the world's most successful models, were replaced with the new 3-cylinder 1020 and 4-cylinder 2020. Conceived as part of the plan for a worldwide tractor, these Dubuque models corrected all the problems of their predecessors.

They had an 8-speed constant-mesh transmission as standard and a vast array of optional equipment including power steering, differential lock, three rear PTO options, and a 1,000-rpm mid PTO.

Following the numerous tractor models built in Mannheim, the 3-cylinder 31-hp 820 and 3-cylinder 46-hp 1520 were offered in 1968. The latter was similar to the European 1120.

The 1020 illustrates the three utility tractor versions, HU high, RU regular and LU low.

Updated Overseas Tractor Models

In 1965, Mannheim introduced the 3-cylinder 310 (32 hp) and 510 (40 hp) and the 4-cylinder 710 (50 hp), all with the new 300 series engines used in the Dubuque 20 series tractors. At the same time the 100 (20 hp) was replaced by the 200 (28 hp), also a 2-cylinder but with an extra reverse gear.

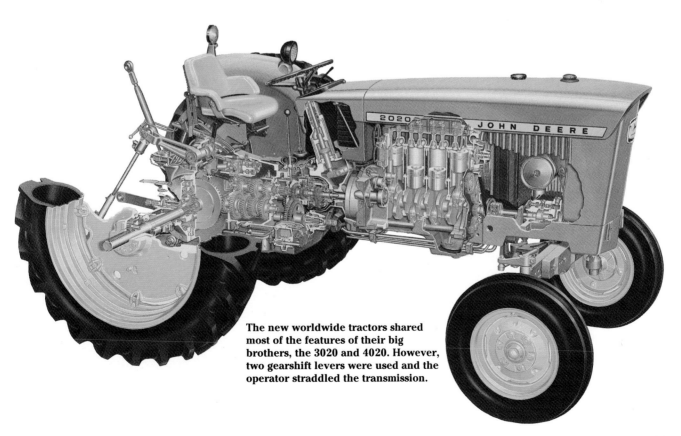

The new worldwide tractors shared most of the features of their big brothers, the 3020 and 4020. However, two gearshift levers were used and the operator straddled the transmission.

The year 1967 saw Mannheim follow the U.S. lead and introduce the 20 series tractors, although in more models to suit this "small farm" market. The 820 (31 hp); 920 (40 hp); 1020, 1020V vineyard and 1020-O orchard (47 hp); 1120 (52 hp); and 2020 and 2020-O (64 hp) were the first models announced.

In Spain Deere & Company gained control of the Getafe operation and changed the name to John Deere Iberica SA. Here it produced the 515, 515V, 717 and 818 tractors.

In June 1968 all the assets of the Lanz company were finally acquired, and a new tractor for Europe and Canada, the 2120, was announced. Zweibrücken replaced the 214 baler series with the new 224 models.

The 6-cylinder 3120, built in Europe, appeared for the first time in 1969 and was instantly popular with the larger farmers.

The beginning of the '70s saw Deere & Company with no less than six proposals to overcome the losses being experienced in overseas operations.

The first of these (the solution finally adopted) was to build all the Dubuque agricultural tractors in Mannheim. Other proposals included the acquisition of tractors built in the U.K. by David Brown or Leyland, either with or without equity involvement, and a tie-up with a Japanese company. But all were dropped when the Deutz deal was initiated. The proposal was to merge with Deutz of Germany, but this was rejected by the company.

At about this time a similar suggestion for a merger with Fiat went much further. In England, for example, Fiat crawler tractors were sold by John Deere dealers for a time, and a full Fiat agency was explored. It was not until January 1972 that negotiations were finally ended. One last suggestion, to close the Mannheim tractor operation, seemed too much like a counsel of despair.

Having explored the possibilities of closing their overseas operations, finding a partner or going it alone, Deere chose the latter option. That decision helped make them the largest and most successful company in the agricultural equipment business today.

For the 1970 season four tractor models were announced in Argentina, the 1420, 2420, 3420 and 4420.

Mexico continued to draw smaller tractors from Mannheim and import the larger ones from the U.S., as did South Africa.

During 1970 Deere & Company and Chamberlain of Australia merged to form Chamberlain John Deere Pty. Ltd. The Chamberlain brothers built their first 40-hp tractors in 1948 with 2-cylinder horizontally opposed kerosene engines. The switch from kerosene to diesel took place in the '50s. In 1970, their Welshpool factory updated their models with the C670, C6100 and Champion 236. The first introductions under the new banner in 1972 were the Chamberlain C456 agricultural and 212 industrial model tractors.

One of the two 1946 Chamberlain 40K prototypes with 2-cylinder horizontally opposed kerosene engine.

In 1952 Chamberlain entered the implement market with an all-welded scarifier suitable for the local conditions, to be followed a year later by a unique and revolutionary disk plow. This machine foreshadowed the design of single-disk roller-bearing plows, which were to become universal in the Australian wheat belt. The need for specialized seeding equipment saw the introduction in 1963 of the combine seeder, and these three lines in modern form are still produced.

Hay Packaged to Flow Like Corn

Hay has been and continues to be a most challenging crop to harvest. It ranks along with corn, soybeans and wheat in acres harvested and production value in the United States. Corn, soybeans and wheat are good cash crops, as the majority of the crop is sold off the farm. They have high value per truckload and are easily conveyed and stored. Conversely, hay is low density and awkward to handle, resulting in more than four-fifths of it being fed on the farm where it is grown.

The shift from loose hay stored in stacks and barns began in the 1940s with the introduction of the automatic twine-tie baler by a competitor. This reduced labor and made a more dense product that was easier to handle, ship and store, but made little change in hay quality.

Some large commercial alfalfa growers in Nebraska had been making hay pellets for a number of years. This was a dense product of high cash value that had higher nutritional benefit and could be handled and stored easily in bags or in bulk. However, pellet production required a high capital investment for the dehydrator and the pellet mill. Drying the fresh cut alfalfa also had high energy costs for fuel.

In the early '50s some university researchers started laboratory work on a machine to make pellets directly in the field. This sounded so much like hay harvesting Utopia that several manufacturers, including Deere, started major projects to develop field pelleters. IH showed an experimental "field pelletizer" to their dealers and the press in July 1958. Lundell started selling a "hay waferer" in 1961.

Spiral bars help the auger uniformly deliver hay in front of the press wheel that pushes the hay out through radial dies.

The 400 hay cuber produces 5 tons of 1-1/4" cubes per hour with its 216-hp diesel engine. The 75 Hi-Dump wagon loads the cubes into a semi-trailer for hauling to another state.

Loaders or flight elevators deliver the cubes in good shape to beef or dairy cattle.

The product development department of John Deere tried many approaches in the laboratory before settling on a design that could be tried in the field. A better understanding of what was required to make a good pellet or cube was gained with field experience in Arizona and California. Alfalfa has an adhesive on the surface of the stems that can be used to cement the cubes together. For this to work, the alfalfa must be dried to about 10-12% moisture content so the stems will not spring back and separate after the cube is completed. To get the adhesive spread throughout the cube, the windrow is sprayed with water and then the hay is chopped for thorough mixing of a variety of particle sizes, much like good concrete.

Complete crushing of the stems and good contact is provided by extrusion pressures of 5,000 to 10,000 pounds per square inch. Compression also produces heat to speed up the action of the natural adhesive. Weeds or grass in the hay prevent making good cubes. Some cubes were made that were so dense they would sink in water. However, the ideal cube for handling and feeding is about two-thirds that density.

While an experimental hay cuber was on field test, the manager of the department visited the site in California. As he returned to the hay-field after lunch on a 100-degrees-plus day, he was shocked to see a steer being butchered by the ranch hands in the open field. He thought to himself that Midwest farmers at least butcher indoors in the middle of winter, when the meat is cooled by nature. However, he soon learned that the impromptu action was needed because the steer had choked on one of the experimental 2" square cubes. Deere paid for the steer, but the lesson learned was much less expensive at this point than it would have been after the cuber was being sold.

Although John Deere hay cubers have made most of the field cubes in the world, the machine was only a partial commercial success.

Self-propelled cubers were sold in limited numbers from 1965 through 1983 in the arid Southwest, West, and certain areas of the Pacific Northwest. In Arizona and California most hay is sold off the farm. It costs less to haul cubes than bales, as trucks can hold twice as many tons. However, most hay producers still prefer balers because of their greater capacity per hour and lower power requirement. Stationary 390 hay cubers were sold over a wider geographic area from 1968-1983.

Considerable further work was done in the '60s to make cubes at moisture levels typical of the north central states, the major market for hay equipment. Good cubes cannot be made above 15% moisture content, even in the laboratory. But higher moisture hay was cubed with some success in the laboratory by compressing parallel stems lengthwise. The cubes were held together mechanically by interlacing, similar to the action of Velcro fasteners. However, the resulting cubes were lower density and ragged, so did not tolerate handling very well.

Roll-Gard Frame Offered to Customers and Industry

John Deere announced the Roll-Gard structure for sale in 1966. The optional canopy helped protect the operator from the sun and rain.

Accidental rear upsets were experienced by some of the early purchasers of tractors. Some models had poorly located drawbar hitch points. Some new users added to the problem of rear upsets of tractors by hooking chains above the drawbar when pulling stumps or other heavy loads. As mowing highway roadsides became common practice, side upsets also occurred. State highway departments started installing their own designs of rollover protective structures (ROPS) on their mowing tractors.

John Deere has had an active safety program for several decades. In 1959 near Moline, a 720 tractor fitted with a roll bar was upset to the rear and to the side multiple times to learn more about rollover protection. This was followed up by further tests at Waterloo on New Generation tractors. These tests determined the location and shape of a roll bar that was able to limit rear or side overturns to about 90 degrees instead of the typical unprotected roll of 180 degrees or more. The strength required of the mast and its attachment to the rear axle were also learned.

The state of Nebraska had and continues to have a very active farm safety program. Waterloo cooperated with the University of Nebraska Tractor Power and Safety Day at Mead, Nebraska, in July 1967. Waterloo furnished two 4020 tractors fitted with dummies for a demonstration of overturn protection. The unprotected tractor rolled a full 180 degrees, crushing the dummy and damaging the tractor instruments and controls so it could not operate. The dummy "survived" the 90-degree roll with the Roll-Gard ROPS-equipped tractor. The tractor suffered minor damage to the hood and canopy but ran as soon as it was rolled upright.

In spite of active promotion by John Deere and various safety groups, customer acceptance of the Roll-Gard structure came slowly and with much effort. Farmer interest in protection from dust, heat and cold became evident in about 1968. Sales increased significantly for factory installed cabs with integral roll protection on 20 series Waterloo tractors.

The real jump came in 1972 with Waterloo's introduction of the 30 series tractors. The majority of these tractors were ordered with Sound-Gard bodies during their first year of sales. This cab had an integral 4-post Roll-Gard structure. Sales of open ROPS also experienced a distinct increase, resulting in three-fourths of the Waterloo 30 series tractors being sold with rollover protection. For the past few years, all John Deere farm tractors sold in the U.S. and Canada have as standard equipment a Roll-Gard structure or a cab with it built in.

John Deere has shown considerable interest in saving the lives of competitors' customers as well as their own. At the time they announced Roll-Gard structures to their customers, they also announced that they would provide free license to other manufacturers for this patented device. Rollover protective structures have probably saved more farmers' lives than any other single safety feature on farm equipment.

There was a precedent for furnishing free licenses for safety devices to other manufacturers. John Deere introduced nylon-bearing tubular PTO shields in 1958 on the Nos. 8 and 9 mowers. Free license to this design was given to other manufacturers. Today, nylon-bearing tubular PTO shields are found on equipment manufactured in the U.S., Canada, Europe and Japan.

Other royalty-free designs of safety features that Deere has offered industry manufacturers include:

A Powr-Gard shield that fully enclosed the PTO driveline.

A tip-up tractor master shield for easier PTO hookup.

A device to avoid tractor movement from bypass starting.

An operator presence system for cotton pickers.

Searching for the Ultimate Transmission

Carmakers are often considered the leaders in introducing new technology. Some cars in 1910 had a mechanical infinitely variable (MIV) transmission consisting of a smooth input pinion driving a large flat disk. Speeds increased continuously as the pinion moved from the outer diameter of the disk toward the center. The function was superior to competition but the drive tended to slip, and wear was a serious problem.

The market leader of the 1920s, the Model T Ford, used a 2-speed planetary transmission. However, in its Model A replacement Ford joined the competition with a manual, multiple-gear fixed-ratio transmission. Sixty years later, this design remains the base transmission in most cars, but the number of speeds has increased from three to four or five.

The first John Deere Model "D" tractors made in 1923 had 2-speed transmissions. The Model "A," introduced in 1933, had four speeds. With rubber tires becoming the norm, two road gears above 5 miles per hour were added to the "A" and "B" in 1941. The 6-speed transmission remained in the "A" and "B" and their successors through 1959.

In the '50s, Waterloo engineers examined many potential tractor transmissions. The 8-speed Syncro-Range transmission on the 3010 and 4010 tractors was a distinct step forward for the industry. The 8-speed Power Shift transmission introduced in 1963 was a further advancement in ease of control.

However, engineers were always searching for the ultimate transmission—one that was easy to control, durable and fuel efficient. Electric transmissions like those used on locomotives were examined. The automotive torque converter offered the desired smoothness but had poor fuel economy, and travel speed varied too much with load.

Hydrostatic transmissions were considered for farm tractors for a number of years because they are infinitely or continuously variable from maximum forward speed through neutral to maximum reverse speed. This ease of control has resulted in the use of this type of transmission on John Deere combines, lawn and garden tractors, and large crawler tractors. Power losses are not too serious for combines or small tractors, because propelling the machine requires only part of the machine's power. In large crawlers, power loss is outweighed by improved maneuverability over short distances, which increases work capacity. However, farm tractors spend much of their time with steady, heavy drawbar loads that require good fuel economy. Thus, this near-ideal transmission has not been adopted on John Deere farm tractors.

The variable-speed V-belt was another continuously variable transmission that had been in use on John Deere combines since the mid-'40s. It

Twin round-cross-section variable-speed chains ran dependably in three experimental 4010 tractors but also required a 3-speed manual transmission in series.

did not offer the user flexibility of the hydrostatic transmission but was lower in cost and had better fuel economy. However, the size required to take the steady high loads of a farm tractor ruled it out.

The size problem would be eliminated if the rubber V-belt could be replaced by a much-stronger smooth steel chain. Wayne H. Worthington, manager of research engineering, found such a design in the Reimers transmission on one of his trips to Germany. Adapting this to an "OX" (4010) tractor began in June 1958.

A clever and unique improvement in the fall of 1960 eliminated manual shifting between 16 mph in forward and .5 mph in reverse. This MIV design provided a 4-to-1 speed variation, with a Reimers-developed single chain having a cross section similar to a V-belt. A power shift at about 4 mph changed the path of the power train so the 4:1 ratio could be used twice. The MIV was

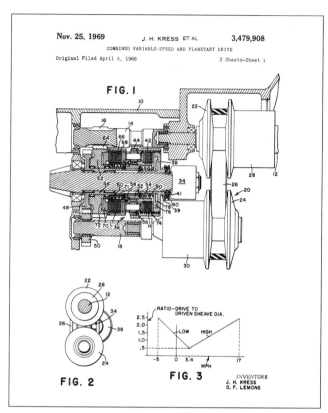

Nov. 25, 1969   J. H. KRESS ET AL   3,479,908
COMBINED VARIABLE-SPEED AND PLANETARY DRIVE
Original Filed April 6, 1966   3 Sheets—Sheet 1

FIG. I

FIG. 2   FIG. 3

INVENTORS
J. H. KRESS
D. F. LEMONS

A patent drawing of the final design, which provided infinitely variable speeds from 17 miles per hour forward to 5 mph reverse, with a single power shift at 3.4 mph.

very easy to operate and had good fuel economy from small losses. In addition to the tractors used by Deere for laboratory and field testing, a 4020 tractor with an experimental MIV transmission was loaned for field evaluation to farmers, one near Vincennes, Indiana, and one near Bottineau, North Dakota. They were as happy with its function as the test engineers had been.

Although the ideal transmission had been achieved functionally, the project was dropped in 1969. Great precision was required in the chain and sheaves to get parts that would have long wear life and not slip. Production tooling cost estimates came in very high for the critical chain drive parts. Estimated sales volumes did not justify this third transmission option.

While the chain-drive continuously variable transmission is still not to be found in any farm tractor, it has been introduced into world markets in a much smaller size in the Subaru Justy

car. Borg-Warner also displayed their CVT (continuously variable transmission) at the SAE meeting in Detroit in February 1990. Thus the car buyer is again able to get the transmission feature he enjoyed 80 years ago, but with improved durability.

New Generation Combines

Two new tractor-drawn combines were introduced in 1964, the small 42 and the large 96. The Long Green Line was certainly getting longer. More significant was the more comfortable operator's platform, including an air-conditioned cab as an option on the self-propelled combines. A redesign of the combine line resulted in larger grain tanks with faster unloading, a new cell-type separating grate behind the concave, and a strengthened separator drive. The new models included corn, rice, edible-bean and grain versions, the latter including hillside and the tractor-drawn models. In 1965, the largest tractor-drawn combine to date, the 106 Hi-Lo, was introduced.

Harvesting equipment was updated during 1968. The combine improvements included new on-the-go cylinder-speed adjustment from 445 to 1,049 rpm as an option on the larger models, dial-type fan-speed adjustment, and increased engine horsepower for the 105 combine to 105 hp. Grain tank extensions, engine cover and all necessary field and highway lights as standard were among the many extra features which made John Deere combines the leaders in their field, and all the new benefits on the self-propelleds were also available on the 96 and 106 pull-type machines. A new pickup reel for use in down and tangled crops, offered in 10' to 20' widths, indicated the shape of things to come. The new Corn Specials gave the closest thing yet to push-button corn harvesting.

The big news for 1970 was the introduction of a completely redesigned line, called the "New Generation." This had been anticipated the previous year with the introduction of the 6601 pull-type. For the first time John Deere abandoned the original layout arrived at with the classic 55 combine and copied by so many competitors. The new combines adopted a layout with an off-

A factory-built cab was standard on the 7700 combine and optional on the smaller models.

set operator's platform and engine alongside.

The four new models, the 3300, 4400, 6600 and 7700, had 25% greater capacity than those they replaced. A 13-bar concave with 2-bar extension gave greater threshing area with a reduction in the power required.

Longer walkers of a new design with greater overhead clearance gave improved grain separation under North American conditions, but proved less effective in the heavy straw conditions found in Europe. A great improvement was the increase in walker crank-shaft throw from 4" to 6" as adopted in the new European line, giving better agitation.

Below these walkers an exclusive multiple-auger system replaced the old raddle and grain pan of the earlier models, ensuring better grain movement, particularly on sloping ground. The opposed-motion chaffer and sieve remained, preventing short straws from bridging between them and giving a balanced operation.

All the usual extras were offered, including straw spreaders and choppers and 3-roller belt

pickups in 88", 110" and 132" widths.

Along with the New Generation combines came the new 40 series corn heads. They had met their design goal of fewer adjustments, fewer lubrication points and less frequent lubrication. Functionally they were superb, and so they set the industry standard for at least the next two decades.

With the increasing requirement for diesel power, the 4400 combine became available in 1972 with a 6-cylinder 329-cu.-in. diesel engine or the 6-cylinder Chevrolet 292-cu.-in. gas model first offered. The 3300 always had the option of both fuels.

Power Front-Wheel Drive and Turbochargers

Hydraulic power front-wheel drive (PFWD) was introduced on the new classic 3020 and 4020 tractors in 1969. They were updated with dry air cleaner, new piston and ring design to reduce oil sludging, and new cylinder block and liners. They were now at 71 and 96 hp. All three new Waterloo models were easily distinguished from the earlier series by their oval mufflers.

A new Power Weight-Transfer hitch was offered to give the same advantages with drawn tools as Load-and-Depth Control gave with integral equipment. The 8-way hydraulic controls and their control console beside the operator's seat gave the tractors flexibility not previously experienced.

John Deere's first turbocharged tractor was announced in late 1968. The 122-hp 4520 shown has optional power front-wheel drive.

A low-priced model, the 4000, with the 4020 engine but only available with Syncro-Range transmission and with other items of equipment as extras instead of standard, extended the tractor line.

The 200-hp X60 tractor program ended in 1968 after obtaining 23,500 pounds drawbar pull in Texas field tests. But customers had a 20-year wait before they could buy a 200-hp John Deere 2-wheel-drive tractor, the 4955. Note the cab, predecessor to the Sound-Gard body.

A "super" 4020, the 115-hp 4320, appeared in 1970 fitted with a Turbo-Built diesel engine, and at the same time the 4520 was replaced with the 135-hp 4620, intercooled as well as turbocharged. Both models could be purchased with the power front-wheel drive option. Syncro-Range transmission was standard on both models but the 4620 could have the Power Shift option.

With the development of its new turbocharged and intercooled engine, the 5020 was uprated to the 175-hp 6030 early in 1972. The previous engine option, which provided 141 PTO horsepower, was offered the following year for this model, the first John Deere tractor to have alternative engines.

John Deere Returns to 4-Wheel Drive

Before John Deere introduced their own new design of articulated 4-wheel-drive tractors, as an interim measure arrangements were made with the Wagner Tractor Co. to market their WA-14 and WA-17 models with slight John Deere styling.

In 1970, the first Deere-designed articulated 4-wheel-drive model since the 8010 and 8020 was announced. The 145-hp 7020, with the same engine as the 4620, used many parts in common with Waterloo row-crop tractors.

In the fall of 1971 a second and more powerful 4-wheel-drive tractor, the 7520 with 175 hp at the PTO, was announced.

The traction provided by the 7020 was useful for ripping and other heavy tillage jobs.

1972–1977 The Sound-Gard Body

The company announces the Generation II series of tractors fitted with Sound-Gard body, a revolutionary new safety cab. Features included underslung pedals, adjustable tilt-telescope steering wheel, seat belts, and an optional radio/stereo tape player.

Generation II Tractors

On August 19, 1972, another giant leap forward occurred in the agricultural machinery world when the Sound-Gard body was introduced. Exceptionally quiet, it was based on an exclusive module design with integral seating, floor, and all the controls, and was attached to the tractor with four rubber mounts. Rollover protection was built in. All four Waterloo models offered the Sound-Gard body or 4-post Roll-Gard structure as options. The cab was an immediate hit with customers, with over three-fourths selecting it on the 4430, the best-selling model.

Once again John Deere had left the competition standing. In standard form the new cab had a pressurizer to keep dust and dirt out; heating and air conditioning were options. The side and rear windows could be swung out if the tractor was not fitted with air conditioning. The large curved windshield had twin wipers, there was also a rear window wiper, and all the glass was tinted and polarized.

Only a limited number chose the power front-wheel drive option available on the 4230, 4430 and 4630 tractors. New-style front suitcase weights were introduced, and rear-wheel weights had built-in handholds for easier and safer installation.

The new 1972 tractors had completely new styling, which took a little getting used to but was quickly accepted by the world's farming community. Included were the 4030 (80 hp), 4230 (100 hp), 4430 (125 hp) and 4630 (150 hp). The 4030 replaced the 3020, with an extra 9 hp from its 6-cylinder 300 series gas or diesel engine. The 4230 replaced the 4020 and continued the diesel or gas option. It was offered in four styles, row-crop, standard, low-profile and Hi-Crop.

The replacement for the 4320, the 4430, was similarly available but without the low-profile version and with a turbocharged diesel engine only. The largest model, the 4630, took over from the 4620; it had an intercooled as well as turbocharged engine and could be purchased in a row-crop or standard model.

All models had the Syncro-Range transmission as standard equipment. Also standard was the new Perma-Clutch, a very durable hydraulically controlled wet-type clutch. Options for the 4030, 4230 and 4430

The Sound-Gard body offered quiet comfort, excellent view, and convenient controls in clean surroundings.

The John Deere Tractor Works, with 2 million square feet, became the jewel in the crown of Deere's 1970s factory expansion program.

included a new 16-speed Quad-Range transmission, which was a blending of the Syncro-Range and a built-in Hi-Lo no-clutch shift. The three larger models were also available with Power Shift, similar to the models replaced.

New articulated 4-wheel-drive tractors completed the 30 series tractor line. The company announced the 175-PTO-hp, 215-engine-hp 8430 and the 225-PTO-hp, 275-engine-hp 8630 for 1975. They were natural replacements for the 7020 and 7520, and they didn't require all new implements.

The 8430 and 8630 tractors were fitted with Sound-Gard bodies and styled to match the other Waterloo tractors.

These new models were marketed worldwide.

As important as the announcement of new machines for 1975, two other matters of long-term significance occurred. A new engine works covering 21 acres was opened near the engineering center in Waterloo. At the same time, land was acquired for a new tractor assembly plant at the "Northeast Site" just outside Waterloo. This would allow the original tractor factory to become the Components Works when the new works was opened in May 1981.

In the spring of 1982, the remodeled and expanded Product Engineering Center at Waterloo was occupied. Worldwide tractor development was concentrated at Waterloo, unlike combine development where design for North America was handled in East Moline while Zweibrücken dealt with European requirements.

Utility Tractors for the '70s

The 2030 had replaced the 2020 in the fall of 1971, and in 1973 three further models were announced, the 35-hp 830 and 45-hp 1530 replacing the 820 and 1520 and the completely new 70-hp 2630 adding a larger model to the line. The two smaller models had 3-cylinder engines; the two larger tractors were 4-cylinder.

The 2030 and its three mates retained the earlier styling and were announced as low-priced tractors, in comparison with the competition.

The next advance in the 40- to 70-hp tractor league was the introduction in 1975 of the 40 series, with the two smaller models (2040 and 2240) sourced from Mannheim and the two larger (2440 and 2640) from Dubuque. In addition to the usual models the 2240 was available in both orchard and vineyard design, the latter with a mere 49.5" overall height and 59" width.

Initially the standard tractors were offered with a 2-post ROPS option, with or without canopy. Shell fenders were standard on all models, with a flat-top alternative for the Dubuque models. There were no less than five PTO options for these two models. The 2040 had a continuous live 540-rpm rear PTO and the 2240s could have an independent hydraulically actuated version as well.

Evenly spaced in size at 40, 50, 60 and 70 PTO hp, the new tractors could be serviced entirely from the right

side of the tractor. The many extras included a wide choice of front and rear wheels, swept or straight front axles, and remote hydraulic outlets.

For 1976 the 2840 6-cylinder 80-hp tractor made its debut. It was introduced in response to farmers' demands for a more powerful tractor with low configuration. Flat-top rear fenders, 540/1000-rpm independent PTO, two double-action remote cylinder outlets, and Category 2

The 2840 had many features in common with the smaller utility models, but the regular transmission had a fully hydraulic Hi-Lo shift with 12 well-spaced forward speeds and six reverse.

The Canadian 3130 included the European suitcase weights with front hitch pin but no European front lights.

hitch with lower-link draft sensing were standard equipment.

Canada has normally shared the same tractor and combine model numbers with the U.S. However, their utility tractors during the time of the 20, 30 and 40 series were the European rather than the U.S. models. Thus as in Europe, the under-80-PTO-hp models remained the 30 series, with the 3-cylinder 40-hp 1030 and 50-hp 1630, the 4-cylinder 60-hp 1830 and 66-hp 2130, and the 6-cylinder 80-hp 3130, but with the new styling of the U.S. 40 series.

The utility tractor line had progressively increased in power by the mid-'70s, so that the smallest agricultural model was now 40 PTO hp. The Horicon Works (Wisconsin) built five tractors from 8 to 19.9 engine hp, and had in fact made half a million units for the consumer products market between 1963 and 1977, but there remained a considerable gap between 19.9 engine hp and 40 PTO hp.

This market had been met in the past by the Ford 8N and similar used tractors. Kubota had stepped in with well-accepted new tractors. In June 1977 a deal was signed with Yanmar, the second largest Japanese tractor manufacturer and John Deere's marketing arm in that country, to produce a line of smaller tractors. A rather complete line of matching implements was available for the 22-hp 850 and the 27-hp 950. Sales since that time have consistently shown that tractors under 40 hp are primarily used for cutting grass. Thus the Yanmar tractors under 40 hp are discussed later under consumer products.

The 1972 Saarbrücken Announcement

European dealers were called to Saarbrücken for the unveiling of the new tractors. The larger tractors built in Mannheim were updated and uprated to become the 2030 at 71 engine hp, 2130 at 79 hp, and 3130 at 97 hp.

Mannheim's smaller tractors remained the 20 series, but special models introduced were the 920-V, 1020-V and 1630-V vineyard, and the 1020-O, 1630-O and 2030-O orchard models.

At this halfway stage in the period covered in this book it is appropriate to survey the position of the factories around the world and their products. The complete Mannheim line of tractors was restyled, without model number change, to conform to the Waterloo series. At the same time the small 20 series vineyard and orchard models became the 930-V, 1030-V and 1030-O, while the larger specialist models also received the new look.

The OPU became available on the 2030 as well as the two larger models. These tractors also had an additional option, hydrostatic front-wheel drive. But the demand in the marketplace was for a mechanical front-wheel drive, so this option was not destined to last long.

The factory at Getafe continued its 30 series tractors,

In 1973, the 59-hp 3-cylinder 1630 was added to the line. All models retained the earlier 20 series styling.

The Sound-Gard Body

An operator's protection unit or OPU was introduced for the new-style 2130 and 3130.

which followed the new-look 30 series from Mannheim but were suitably adapted for Spanish requirements.

In Mexico the 4235 and 4435 tractors gave them new top-of-the-line models. Mexico added two smaller tractors to their 35 series, the 2535 and 2735 models.

Argentina announced the 30 series to replace their 20 series. The new models were the 2330, 2530, 2730, 3330, 3530, 4530, 4730 and 4930.

Australia Introduces "Deere Influenced" Series

In Australia John Deere's influence on the Chamberlain line was apparent in the new Sedan series announced in 1975. The 3380 (68 hp at the PTO) used a 4-cylinder 239-cu.-in. engine. The 4080 (85 hp), 4280 (98 hp), and 4480 (119 hp) were all 6-cylinder, the 4080 with 329 cu.in. and the larger two with 359 cu.in. The 4480 was turbocharged and all four engines were supplied from the Saran factory.

The two smaller models could be purchased with fixed or adjustable tread ("track" in Australia), the latter when supplied with Category 2 3-point hitch, but the majority of tractors "down under" were used with drawn equipment using the remote hydraulic outlets supplied.

At the center of the Sedan concept was a heated and air-conditioned cab that provided the comfort of the family sedan. This was a rub-

The 4480 featured the now-standard John Deere front-mounted fuel tank, with side screens for the radiator air intake. Painted Chamberlain yellow and black, the 80 series tractors looked like industrial versions of the North American models.

ber-mounted isolated capsule, with a fully tested 6-post protective frame built in. With its tinted glass, hinged side and rear windows, and deluxe seat with seat belts, it was the Australian equivalent of the U.S. Sound-Gard body. It was standard on all models, although the 3380 could have the ROPS frame as an option.

The transmissions used on all models, again of obvious John Deere origin, were the constant-mesh type with collar-shift Hi-Lo giving 12 forward (up to 21.6 mph) and four reverse speeds. The whole series was built with the necessary weight and strength to cope with the tough conditions of the Australia-New Zealand farming scene, from the cast steel nose to the heavy-duty drawbar.

Cross-Shaker on European 900 Series Combines

European dealers called to Saarbrücken for announcement of the new tractors also saw a new line of combines, the 900 series. Newly styled but still with the basic 55 combine layout of center operator's platform and engine behind the grain tank, the new models featured a cross-shaker with slowly rotating oscillating tines above special-

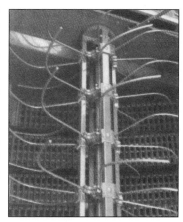

The cross-shaker was a further attempt to cope with heavy straw conditions. It has proved a permanent and most successful additional grain-saving feature of the European models.

ly designed straw walkers. The new line consisted of the 930, 940, 950, 960 and 970, and all had Saran-built diesel engines.

The smallest Zweibrücken combines, the 330 and 930, were replaced by two new models, the 925 and 935, for the small-acreage European farmer. Later, Zweibrücken announced a new series of combines, the 945, 955, 965 and 975. Restyled and fitted with larger grain tank capacity and stronger diesel engines from Saran, they

The 975 combine from Zweibrücken, Germany, was sold in Argentina and several other export markets.

retained the 900 series cross-shakers in all but the two smaller models, and enhanced the European combines' reputation.

Europe saw the 965-H combine announced for those with enough sloping ground to justify its level-thresher advantage.

Late in 1977 the 940 combine was replaced by a new model, the 952, which did not have the cross-shaker straw-separating system of the larger models.

The Large "Invisible" Tractor Fuel Tank

During the rush spring-tillage season farmers would like to be able to run at full load at least 12 hours without stopping to refuel. They often judge fuel economy by frequency of refueling rather than the amount of work done per gallon of fuel.

The front-mounted fuel tank of the 20 and 30 series tractors offered a cool location and was well accepted by users. However, at full load some utility tractors would run only six hours and some row-crop tractors only eight hours without refueling. (Since the engine is normally not fully loaded, the tractors actually run longer than this in the field between refills.)

Three changes were in progress that demanded another look at the location and size of the fuel tank. Radiators needed to be increased in frontal area, to reduce fan noise and fan power consumption. The proposed 4840 had a distinct increase in size (to 180 horsepower) and appetite for fuel. If this appetite were to be satisfied for 12 hours at full load, it would require as much fuel tank capacity as two 55-gallon drums.

Other tractor fuel tanks had been located over the engine, behind the engine, behind the seat or below the platform on either side of the transmission. None of these locations offered the capacity desired while still providing unobstructed visibility, crop clearance, and good implement attachment.

Danny Gleeson sketched a solution that met all of these requirements. The location? In the drive wheel tires. A single 18.4-38 tire, popular on row-crop tractors, has a liquid capacity of 100 gallons.

A fuel system using a drive tire as a fuel tank was designed and field tested. An engine-driven pump forced fuel into the tire during refueling and maintained air pressure in the tire as fuel was used. Three dipper tubes with check valves in the tire returned the fuel to a small surge tank. The system proved to be functional in the field. However, there were some problems in bonding an inner liner in the tubeless tire that was compatible with diesel fuel. The slight leaks resulted in some loss of fuel and air, as well as permitting the tire to slip on the rim. Over time, these leaks caused blisters and separation of the tire sidewalls.

Since the technology of materials required for diesel fuel hoses was well developed, it can be assumed that adequate bonding of a fuel-resistant liner could have been developed. However, there was also a concern over some increase in cost of this fuel system, so the pro-

The quarter-scale mockup of the X61 gas-turbine tractor eliminated the steering wheel and rockshaft. The real tractor was never built.

ject was eventually dropped. The 4840 simply adopted a larger front-mounted fuel tank.

A Gas-Turbine Tractor

Gas turbines were a popular subject in the '60s. Jet aircraft were well proven and the Fairchild F-27 helped establish the propjet. If gas turbines could power propellers, why couldn't they power land vehicles as well?

Although Parnelli Jones's gas-turbine car in the Indianapolis 500 caught the imagination of the public, industry engineers were more interested in the work of Chrysler, Ford, and General Motors Detroit Diesel. Several John Deere engineers had an opportunity to drive one of the many Chrysler gas-turbine demonstration cars. It was a bit sluggish in town but impressive at highway speeds. Ford had several gas-turbine trucks out on road tests. When GM had gained enough experience, they announced plans to build gas-turbine engines for their buses.

John Deere engineers had gained some experience with high-speed turbines in the development of the turbocharged 4520. The engineering research department in Moline installed a purchased gas-turbine engine in a scraper in 1967. A Caterpillar J621 scraper was chosen for the demonstration unit, as it was larger than any made by John Deere. The results were promising enough that Bernie Poore transferred with the project to Waterloo in 1969. By November 1969, Waterloo had a 5020 operating with an Air Research gas-turbine engine.

The appeal of the gas turbine was its smooth, quiet operation, low emissions, high power in a small package, low maintenance and long life. These advantages led to planning a family of three engine sizes. Engineers concentrated on the middle-size 250-hp unit, expecting the most popular farm tractor to require a 200-250 horsepower engine in 1975-1980. The best-selling John Deere tractor had almost doubled in power from the 50-hp 730 in 1960 to the 96-hp 4020 in 1970. However, the best-selling John Deere tractor in 1990 was the 4455 at 140 PTO hp.

The first John Deere-designed T250R gas-turbine engine was built in June 1970 and the second was completed by October. Development work concentrated on reducing the three main disadvantages of the gas turbine: high cost, no torque rise as loads decrease engine speed, and poor fuel efficiency.

Probably the greatest progress was made in reducing costs. A simple single-shaft design was chosen, with a single-stage compressor and sleeve bearings. An investment-cast nickel alloy turbine rotor was used, at a fraction of the cost of the machined rotors in aircraft engines. The rotor operated satisfactorily at speeds of over 60,000 rpm.

Some current John Deere diesel engines increase their torque by at least one-third under variable load conditions typical of farming. With no torque rise available from the gas turbine engine, a special transmission with an automatic control would be required. The first hope was the mechanical infinitely variable transmission previously described. When it was dropped an attempt was made to develop a suitable low-loss hydrostatic transmission.

It was probably the energy crisis of 1973 that clinched the demise of the gas turbine, due to its overall poor fuel efficiency. John Deere had improved fuel efficiency some by safely increasing turbine inlet temperatures. However, they were never able to get a dependable rotary regenerator, the other main avenue to fuel savings. Having to package the engine with a less efficient transmission compounded the problem. And during this period, diesel engine technology made great strides with the development of turbocharging and intercooling. Thus the project faded away in 1973,

Deere & Company purchased the Curtiss-Wright facility and the North American rights to the Wankel rotary engine in 1984. Since then several sizes of rotary engines have been further developed and tested. Some of these run on a variety of fuels. All offer reduced size and vibration, compared with diesel engines.

because improvements in the gas turbine had not caught up with those in diesel engines.

Increased Forage Harvesting Capacity

Harvesting capacity is frequently determined by how well the crop feeds into the machine in difficult conditions. The Ottumwa Works (Iowa) greatly reduced feeding problems in row crops on their 34 and 38 forage harvesters with the introduction of rubber gathering belts. The belts were flexible enough to handle all sizes of stalks and feed them uniformly into the cutter-head without bunching or plugging. For leaning and tangled stalks, a spinning power corner helped feed the crop without hairpinning.

The 5400 self-propelled forage harvester with 1510 high-dump wagon offered the large-acreage operator high daily capacity. Hydrostatic rear-wheel assist kept the equipment running in wet fields.

The next step-up in harvesting capacity came from increased size and power with the introduction in 1972 of the 5200 and 5400 self-propelled forage harvesters. A mower bar, windrow pickup, and 2-, 3- and 4-row ear corn heads and row-crop heads permitted these harvesters to work in a variety of crops and conditions.

Several improvements in forage harvesters came in 1979 with the Dura-Drum cutterhead.

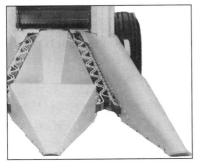

The solid drum and staggered mounting of the knives moved the cut material more positively to the cross auger. Knife backs were coated with tungsten carbide for longer wear between sharpenings,

Patented gathering belts for feeding row crops into forage harvesters were an industry first.

which now required only 10 minutes. Shear bar adjustment was simplified to 2 minutes' time and the use of one wrench. Several details including radial-arc feedrolls contributed to a more efficient, more uniform cut.

The Dura-Drum cutterhead uses 36 or more knife segments to reduce cost of replacing knives damaged by rocks or metal.

A Truck Designed for Farming

The transport of harvested crops from the field has been and continues to be one of the major bottlenecks in farming. This dilemma has led to the evolution of a variety of equipment and methods used. The problem is accentuated by a continual growth in the capacity of harvesting equipment.

The 50- to 60-bushel grain box on a steel-wheeled wagon pulled by horses was adequate to haul grain from the threshing machines of the 1930s. Its flare-box replacement on rubber-tired gear pulled by a tractor could hold three dumps from John Deere's 12A combine, its best seller in 1950. The 55 combine had a grain tank that grew from 45 to 55 to 65 bushels, but a barge box was still a reasonable solution.

The New Generation 7700 combine introduced in 1970 compounded the grain-handling problem with its 129-bushel grain tank. Wheat farmers could normally unload on the go into regular farm trucks for transport directly to the elevator. Handling of corn and soybeans took a more diverse route due to the frequently muddy fields. Combine grain tanks fill more rapidly in corn, partly because corn yields are more than double those of wheat and beans. Gravity boxes pulled by farm tractors were low in cost, could take one dump in the field on the go if not too muddy, and were satisfactory to haul to on-farm drying or storage. Grain carts behind heavier farm tractors could take two dumps on the go and then transport the grain to the farmstead or transfer it rapidly to a truck on the roadside for hauling to the elevator. Farm trucks could be used for unloading on the go if the ground was solid, but more often they had to be parked at the road for unloading at the end of the row. Use of farm trucks in the field resulted in some drivetrain failures due to the severe service.

Wilbur Davis and other Waterloo engineers with current and past farm experience thought there should be some way to use tractor components to design a transport vehicle specifically for farming. In muddy fields it should have the flotation and mobility of farm tractors or harvesting machines. On highways it should be able to operate up to 50 miles per hour, like farm trucks.

The first 4-wheel-drive unit with 18-inch-wide flotation tires was shown to management in the fall of 1970. During its first year, the AMT (agricul-

The agricultural materials transport used a 4630 engine and modified Power Shift transmission to provide a wide range of field and road speeds. Many other production components were used for the rear axle, hydraulics, PTO and cab.

Six AMTs were placed in six sales branches, spread from coast to coast of the U.S., to get experience in a wide variety of crops and field conditions.

tural materials transport) was used to haul harvested grain and forage from the field. It was also used to haul and spread dry fertilizer, anhydrous ammonia, and manure. In the winter it was used for snow removal.

The first AMT tested out well enough to show it to John Deere dealers when the 30 series tractors

were introduced. Six more units were completed in the fall of 1972. The four units with 4-wheel drive had a net load capacity of 20,000 pounds each, with the load distributed equally on all four wheels. The other two units had six wheels and 32,000-pound net load capacity.

A year later, these six machines had accumulated over 3200 hours of use on farms. Two-thirds of the usage was to transport the following harvested crops from the field: corn, forage, sugarcane, sweet corn, lettuce, potatoes and small grain. The next largest use was to haul and spread manure and herbicides. Some work was done hauling feed in feedlots. Disking was the only plain drawbar work in the field.

No additional machines were made in 1974 due to a serious concern over the potential selling price of the machine. It appeared that sales of 1,000 AMTs per year would require not only the base farm market but also selling to the military and other users requiring high mobility. By now the price of the 6-wheel unit had risen to that of a 4-wheel-drive tractor, putting it out of reach of the mass farm market.

Lower cost alternatives were tried that used either large 2-wheel-drive or 4-wheel-drive tractors to pull conventional highway semi-trailers. The project was dropped because it was difficult

41

to develop any new alternative for grain handling that made economic sense to enough farmers to make it profitable for John Deere.

The major competitor for hauling grain on the highway remains the farm truck, which is often purchased used. Combine grain tank capacities have continued to rise, to 240 bushels for the 9600 Maximizer combine introduced in 1989. The John Deere 500 grain cart, with its 500-bushel capacity, will hold two dumps from the largest of combines and can be pulled with the larger 2-wheel-drive tractors. Competitive grain carts are available with double this capacity but require 4-wheel-drive tractors to pull them. Gravity boxes remain a popular alternative for local hauling.

Rotary Hoe Sets New Standards

Rotary hoes are used to break crusted soil and to uproot small grass and weeds in row crops like corn and soybeans. John Deere's horse-drawn hoes were 7 feet wide to cover two rows. Sixteen-tine malleable-iron hoe wheels were staggered on two long axles to give 3-inch spacing. Individual heat-treated steel tines were introduced in 1938 for longer wear and less bending.

This basic design remained in the tractor units of the '60s except that each pair of axles was shortened to cover only a single row for greater flexibility. The biggest functional improvement of the tractor rotary hoes was their much higher operating speeds, which resulted in considerably better shattering of crusts and killing of weeds and grass. The buyer had the choice of drawn units, drawn units with lift wheels, or integral 3-point hitch models.

Tractor power and resulting operating speeds continued to rise. The 42"-wide individual row units lacked flexibility when hitting rocks and did not conform to crops grown on beds. Des Moines Works (Iowa) engineers designed a better solution that was introduced on the 400 series rotary hoes in 1972. The time-proven 16-steel-tine wheels, mounted in a staggered fashion for trash handling, were retained. Each wheel was mounted on a sealed ball bearing and each pair of hoe

Spring-loaded tandem wheels ride easily over rocks and stumps at today's high tractor speeds.

wheels was offset on a walking beam attached to a spring-loaded shank. This design gave constant down-pressure on varying ground contours and provided spring-trip action when rocks or stumps were hit. Shanks were mounted on either rigid or folding box-beam toolbars. This flexible design allowed the widest unit to work equally well for 16 narrow (30") or 12 wide (40") rows.

This greatly improved design increased industry sales and improved John Deere's market share. The design was so good that most competitive units soon had a similar appearance. However, each of their hoe wheels had to be mounted on a separate shank to avoid infringing Deere & Company's patented design.

Max-Emerge Planters

John Deere entered the '60s as the market leader in planters. The 494 and 694 would check corn at 5 miles per hour while the 495 and 695 would hill-drop or drill at 7 mph. John Deere's planter reputation had been established by the 2-row horse-drawn 999, which reigned for over three decades. Sales leadership came with the tractor-drawn 290 and 490 planters, which dominated the John Deere line from the mid-'40s to the mid-'50s.

John Deere introduced the first plateless planter in 1968. The 1200 series planters which replaced the 494-695 planters were not as well liked by farmers, so market share declined. IH followed with a major improvement in their planters, a plateless design known as a Cyclo-Planter. Design of Deere's replacement for the

1200 was under way but production was still to come. Market share was eroding. What could be done?

William A. Hewitt had recognized the need for all Deere managers to understand the corporate principles that had evolved and stood the test of time for more than a century when he published a series of two dozen "Green Bulletins" in 1964. A business principle that applied directly to this case of declining market share for planters had been stated in one of the "Green Bulletins" from the beginning. It quoted a 1911 sales branch policy:

"Do not talk about competitors' goods or mention the names of competitors' goods.

Whenever you talk competitive goods you are advertising them instead of your own."

Some people felt that an exception to the above rule was justified in the crisis caused by John Deere's major competitor having a recognizably newer planter design. To counteract the situation in the Corn Belt, radio advertising was used to compare the two planters. A follow-up survey of several John Deere and some IH dealers in Iowa revealed that the negative advertising had increased sales for IH two ways. First, it gave the Cyclo-Planter name wider exposure, and second, it indicated that John Deere was running scared.

The 7000 drawn and 7100 integral planters introduced in 1974 so advanced planter technology that the slide in market share quickly reversed. Since 1977 John Deere has sold the majority of all planters in the U.S. and Canadian markets. Traditionally, John Deere market shares have been highest on those products where quality of function has the most direct effect on the farmer's income. Thus, farmers are willing to pay a premium to get John Deere quality planters, combines, and cotton pickers.

What made these planters so much better? Engineers used a basic approach, focusing on optimum seed environment to get maximum germination and emergence for important crops like corn, soybeans, cotton and sorghum. Agronomists at the Technical Center ran field tests as well as laboratory soil-bin and growth-

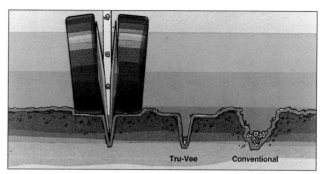

Germination is improved in marginal moisture conditions by minimizing soil disturbance and keeping dry soil out of the furrow.

chamber tests. For best emergence, the soil should be firmed around the seed but left loose above it for minimum resistance and to reduce crusting. The result was the now well-known Tru-Vee openers with angled closing wheels.

Farmers demand good germination and emergence because the influence on yields is obvious and well proven. They also want uniform spacing, although it has a lesser influence on yields. The finger pickup was the most popular seed meter for corn on the 7000 and 7100 planters. The feedcup or traditional plate meters were used for other seeds. The accuracy of the original metering is maintained with a curved seed tube that "dead drops" the seed.

Two other trends in planter design became quite pronounced in the '70s and '80s. Reduced

John Deere sold more than a million of the original Max-Emerge planter row units.

tillage in a variety of forms meant that planters needed to work in more trash and probably harder soil conditions. Going from the traditional wide rows (38" or 40") to narrow rows (30") showed a distinct yield increase for soybeans and some increase for corn.

The company helped ensure its continued position in the market in 1986 by the introduction of the 7200 drawn and 7300 integral Max-Emerge 2 planters. These planters feature a vacuum meter, suitable for corn, soybeans, cotton and sorghum, as well as the previous three meter options. Row spacings as narrow as 15" are available for soybeans and other crops.

Row-Crop Head Reduces Soybean Losses

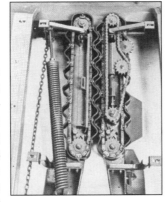

Soybean production in the U.S. doubled in the '70s, sparking increased interest in specialized harvesting equipment. Soybeans are notorious for high harvesting losses due to their low-growing pods and their ease of shattering. They were originally combined with conventional grain

The stalk is gripped just prior to being cut off by a rotary knife, to minimize shattering losses in the Row-Crop head.

heads used for wheat and other small grains. Pickup reels gathered more pods and provided more positive feeding with less shattering. Flexible, floating cutterbars further reduced losses by allowing the platform to fit ground contours and get more of the low-growing beans.

The exclusive 50 series Row-Crop head introduced by John Deere in 1975 further reduced losses and remains the minimum-loss way to harvest soybeans. The Row-Crop head saves about 1.5 bushels per acre more than a flexible platform and 3 bushels more than a rigid platform. The basic technology for this header was derived from previous experience with the rubber gathering belt on Ottumwa's forage harvesters. The

Hydrostatic rear-wheel drive keeps combines moving in muddy rice, corn and soybean fields.

Row-Crop head is also popular in sorghum, sunflower and millet.

For 1976 the North American combines had a number of new attachments added, including the 200 series platforms with flexible or rigid cutterbar, a new pickup platform with 88", 110" or 132" 3-roller belt pickup with hydrostatic drive, straw choppers, a deluxe Personal Posture seat option, and full field lighting.

The 1977 season brought a new 7701 pull-type combine, hydrostatic power rear-wheel drive for the 6600 and 7700, and increased engine power for the 4400, 6600 and SideHill 6600 models.

Combining Corn on the Level

Many John Deere engineers retain sufficient roots on the farm to feel the necessity that is the "mother of invention." On a frosty morning in November 1970, Murray Forth was deer hunting on his rolling farm near Morrison, Illinois, with his friend, Bud Bichel. Both were jolted by the amount of shelled corn left in neat windrows from each pass of the combine in a freshly harvested sloping field. Something had to be done! Fate could not have selected a better pair to do that "something," since Forth was product concepts manager at the Technical Center and Bichel was senior division engineer on combines at the Harvester Works.

A center pivot pin on the feeder house permitted headers to remain parallel to the ground while the combine corrected for slopes up to 18%. Six rubber paddles replaced the conventional conveyor chain in the feeder house.

Bichel knew that self-leveling hillside combines, sold in the Pacific Northwest for wheat, would not have had these high grain losses out the back, because the separator would have been operating on the level. Forth knew the majority of Corn Belt farmers could not afford the added cost of the hillside feature. Together, they reasoned that slopes in the Midwest were less than half those in the Palouse area, so self-leveling might be done more simply and at lower cost.

The concept developed at the Technical Center proved itself in field tests. One experimental combine was placed on a farm near Dubuque, Iowa, and another on a farm near Wheatland, Indiana, for two harvest seasons. Test results and farmer response were sufficiently encouraging that the project, along with engineer Kent Cornish, was transferred to Harvester

Works. The SideHill 6600 combine was introduced in 1975.

The 6600 combine was chosen for offering this SideHill option because it was the best selling size in corn. Headers had to be modified also. Corn heads and Row-Crop heads were offered in either 4-row wide or 6-row narrow sizes. Grain platforms of 13, 15 and 18 feet were made available.

The economic reason for buying the SideHill 6600 combine was a combination of saving time and saving grain. On side slopes the level operation of the combine permitted faster travel with reduced losses because the crop was spread uniformly across the cleaning shoe instead of sliding to the downhill side. The greatest payoff was in corn, but soybeans and wheat also showed considerable improvement.

Jim Graf, the Dubuque test farmer, said after running an experimental unit two years, "We went about 25 percent faster and saved more hill crops with a test John Deere SideHill combine. I think that for the extra cost, this combine might pay for itself in two years' time with the amount of crop it saves in the hills."

Two deer hunters' dreams had been fulfilled!

An additional fringe benefit of the SideHill design was comfort of the operator, who remained level during long harvest days. Customers also discovered that the equal weight balance on the two drive wheels resulted in better traction on slick slopes.

No More Tricycle Tractors Made by Deere

The passing of the tricycle tractor was less abrupt and traumatic than the passing of the 2-cylinder engine. Part of the reason for this was the customer shift in preference from the tricycle to the wide-front-end tractor.

There is still some reason to shed a tear, because it was the tricycle Models "A" and "B," introduced in the mid-'30s, that opened the mass market for John Deere tractors in the Corn Belt. Up to that time, tractors had primarily been used for tillage, but this configuration permitted culti-

vation of row crops. In the '40s, mounted 2-row corn pickers became popular and they demanded tricycle front ends also.

As the two machines that led to the birth and growth of tricycle tractors were phased out by farmer choice, the need for tricycle tractors declined. The popular Ford-Ferguson tractor and its early successors offered only a wide adjustable front axle. Their row-crop cultivator in the rear on the 3-point hitch was much simpler to mount. Customers and engineers of John Deere and other major tractor lines were very skeptical of this design. Some even said that looking back at a cultivator to stay on the row was as haz-

The experimental 60-hp 2120 tricycle tractor was shown at the 1968 Waterloo product review meeting as a lighter weight, lower cost alternative to the 2520 tricycle tractor.

ardous as the action of Lot's wife (Genesis 19), who turned into a pillar of salt when she looked back. They lacked faith in the row guide mounted on the front axle and certainly did not foresee that this design would overtake front-mounted cultivators in popularity.

John Deere engineers reluctantly provided two choices to the customer who did not require a tricycle for a mounted picker and did prefer a wide front end. The first alternative was rear-mounted cultivators. The second was a modification of the traditional Quik-Tatch front-mounted cultivator. The pivots were moved out and the gangs located to clear the front wheels. Rolling shields and wider set-tings between the shovels next to the row made clogging with trash less likely and thus reduced the need to watch each row, making rear mounting more practical. The lower cost and faster attachment of rear-mounted cultivators caused their sales to pass those of front-mounted cultivators for John Deere in 1964.

John Deere was the industry leader in offering corn heads for combines in 1954. At first they were limited to two rows and thus offered little more capacity than mounted pickers. However, customers were quick to recognize that combines could operate in softer, muddier conditions than tractor-mounted pickers. Combines did not have their weight concentrated over two small front wheels and were not pulling a wagon behind.

The popularity of corn heads took another jump as capacity was increased with 4-row and larger units. U.S. shipments of corn heads doubled from 1962 to 1964, when they reached over 17,000 units. This was the first time that shipments of corn heads exceeded those of mounted corn pickers. Sales of corn heads in the U.S. and Canada continued to rise and peaked at over 28,000 units in 1973.

Tractor engineers agonized each time they dropped the tricycle option from a tractor model. The last tricycle tractor from Dubuque rolled off the assembly line in 1965 when the

Owner-operator Myra Thompson still prefers the simplicity of the tricycle design in her new tractor. Rider Marilyn Fish enjoys her 1/32 scale current style tricycle tractor. All 1/16 and 1/64 scale models of current John Deere tractors have wide front axles.

2010 was replaced by the 2020. The worldwide 1020 and 2020 designs were too low to permit wheels directly under the front end. Europe did not need tricycles and their sales were on a decline in the U.S. and Canada. However, to meet demands from marketing, the 2510 was introduced the same year. It had a 2020 engine in a 3020 chassis.

Single-wheel, dual-wheel, and Roll-O-Matic tricycle front ends were offered on the 4030 and 4230 tractors. On the best-selling 4430, only the dual front wheels offered enough tire capacity for its weight. No tricycle option was available on the 4630. The last of the tricycle tractors from Waterloo came in 1977 with the switch from the 30 to the 40 series. It is estimated that John Deere built more than one million tricycle tractors in the 50 years they were produced.

For tricycle tractor fans there is some solace still at your John Deere dealership. Since their beginning in 1945, the ERTL toy company has made several models of John Deere pedal tractors. All of those based on farm tractors have been tricycles. The current model outsells John Deere's most popular farm tractor, the 4455.

1977–1982 The Iron Horses

The entire 40 series of tractors from Waterloo gave "more horses from more iron," enabling them to truly live up to their name. As if to emphasize the advances made, a new 2-year warranty was introduced on the engines and the heavier powertrains.

Engine Displacement Increased

Announced in the fall of 1977 for the following season, a completely new line of five 90- to 180-hp tractors appeared from Waterloo. With the exception of the 4040, all the tractors had new engines of 466-cu.-in. displacement. The 4040 had the naturally aspirated 404-cu.-in. engine used in the 4230, but developing only 90 PTO hp.

The larger 466-CID diesels ran at 2200 rpm, with the 4240's naturally aspirated version giving 110 hp, the 4440's turbocharged unit giving 130 hp, and the 4640's (both turbocharged and inter-cooled) giving 155 hp. Farmers had asked for a Sound-Gard-bodied 6030. The new top-of-the-line 4840, with 180 hp (and Power Shift transmission as standard), provided the answer by using the same engine as the 8430.

The new engines retained all the best features of the previous models but had a host of new ones, including bigger pistons, rods, rings and

The 4240 and its four companions featured larger fuel tanks for longer work days, and better cooling from larger radiators, fans and water pumps.

wrist pins. They had new fuel pumps, distributor-type on the two smaller tractors and in-line on the three larger, plus bigger-capacity alternators on all models.

The Waterloo Dynacart was used to check the drawbar muscle of the new 180-hp 4840 tractor.

The popular 16-speed Quad-Range transmission was standard on 4040 through 4640 models. Two additional models were announced at the same time as the row-crops, both the 4240 and 4440 being offered as Hi-Crop models with 4-post Roll-Gard structure as standard.

Hydraulic capacity was greatly increased, both at the 3-point hitch and external services, with a 17% increase in the transmission charge pump capacity allowing the use of bigger loaders and other high-capacity implements.

By lengthening the wheelbase and fitting a new Personal-Posture seat, a new standard of comfort was achieved. More insulation and padding in the Sound-Gard body made it even quieter.

A John Deere engineer once described a tractor engine as "a honeycomb casting in which you keep fuel, oil, air, and exhaust separate…set the whole thing on fire and plow a field with it." The new models certainly set the tractor market on fire.

Under-90-hp 40 Series Tractors Improved

In the fall of 1979 the company introduced an updated series of tractors in the 40- to 80-hp sizes without changing the model numbers. New options included mechanical front-wheel drive and a top-shaft-synchronized (TSS) transmission

The 2940 had enough power to be the main tractor, and the maneuverability to work well with the 347 baler and other hay tools on cattle farms.

For the 1982 season the 2940 in the U.S. and the 3140 in Canada were offered with Sound-Gard body.

for the 2040 and 2240 models built in Mannheim.

The 8-speed collar-shift transmissions remained as standard on all the 3- and 4-cylinder models. Hi-Lo was another option for the 2240, 2440 and 2640, and for these models yet another choice was available for back-and-forth jobs, a hydraulic direction reverser.

Since it too was made in Germany, the 2940 had a similar MFWD option but with the TSS transmission as standard with Hi-Lo shift in addition, giving 16 forward speeds. All models had a new easily adjusted and more comfortable seat, an electronic instrument panel, and a new park brake with audible and visual warning systems.

The engines had beefed-up oil pumps and specially coated piston rings for longer life. The 3-cylinder models had longer-life injection nozzles for better fuel economy, and 24% larger fuel tanks. The 2040 engine displacement was increased from 164 to 179 cu.in.

The 2440 and 2640, which continued to be built in Dubuque, had improved engine balancer shaft bearings for longer life. The 2940 also had its engine displacement increased from 329 to 359 cu.in. and the engine block strengthened. A new engine-oil cooler and full-width radiator added to increased performance and reliability.

With the engine and power train now warranted for 1500 hours or two years on all models and with more than 20% increase in the 2440 and 2640's rockshaft capacity, the new tractors represented a further advance in the medium-size tractor's profile of performance.

New Medium-Size Tractors in Europe

While the Mannheim models 930 through 3130 were continued for 1978, two additional tractors joined the line: the 830 with 32 PTO hp and the 3030, another 6-cylinder model that was effectively a 3130 derated. A hydrostatic front-wheel drive option was available for the new 3030 as well as the 2130 and 3130.

Mechanical front-wheel drive was offered for the 1030, 1130, 1630 and 2030. Variations were the 1030-V and 1630-V vineyard, the 1030-O, 1630-O and 2030-O orchard, and the Universal 1630 and 2030 tractors with higher crop clearance.

The 1640 and other European models had front lights and rear fenders that were different from comparable tractor sizes sold in the U.S.

At the same time that the company was improving the medium-size range of tractors for the U.S. market, the tractor line built in Mannheim was similarly updated. There the 30 series also gave way to the 40 series. The new models announced for the 1980 season were the 3-cylinder 840, 940, 1040, 1140 and 1140 narrow; 4-cylinder 1640, 2040 and 2140; and 6-cylinder 3040 and 3140. All 1040 through 3140 models were offered with the OPU cab introduced five years previously on the larger models.

These new 1980 tractors were labeled the "Schedule Masters," since they were designed to finish work faster and do more in a given time. New engine features, Hi-Lo option from the smallest model up, and redesigned wet-disk brakes were among the new features of the 40 series.

More significant was the mechanical front-wheel drive option on the 1640 and larger models, a development that would have far-reaching repercussions. This gave 4-wheel braking, urgently needed on Europe's steep hills but not available with the previous hydraulic drive assist.

A new factory at Bruchsal near Mannheim, Germany, came on stream in the summer of 1981, manufacturing Sound-Gard bodies for the European tractor line from the 2040 up. Additional tractors built or assembled in Mannheim were the 2040S, 4040S and 4240S. The three S series tractors each gave about 5 hp extra.

The two larger tractors could be ordered with the new Caster/Action mechanical front-wheel drive option, which had the unique ability to lean at up to 13 degrees to allow tighter turns.

Other Overseas Tractor Developments

A new tractor factory was opened in Venezuela in 1978 following a pilot assembly of some 3,200 Waterloo tractors, but by 1980 the operation had to be closed due to an economic downturn in the country. In addition John Deere sold $1,000,000 worth of equipment to China at this time.

The year 1978 also saw the Argentina factory virtually closed due to a farming depression. Both Argentina and Australia introduced the 40 series tractors from Mannheim for their lower power requirements, the 2140 through 3140 in Argentina and the 1040 and 3140 in Australia. The Japanese-built 850 and 950 tractors were also added in Australia. In Argentina the 4040 went into production in 1982 for a couple of years.

The major change for that year occurred in South Africa; the government had decreed in the late '70s that the diesel engines for tractors should be built in that country. As a result the Atlantis Diesel Engine Company (ADE) was formed, a subsidiary of Perkins. In February 1982 the new ADE assembly facility was commissioned and John Deere announced the new 41 series tractors, which were the Mannheim 40 series with these locally produced engines. Design work was primarily done by Mannheim engineers, with some input from engineers at the Nigel Works in South Africa. The tractors ranged from the 4-cylinder 41-PTO-hp 1641 to the 6-cylinder 66-hp 3141.

New 4-Wheel-Drives

The large 4-wheel-drive 30 series tractors were replaced by the 8440 and 8640 with similar power output. A number of detail improvements arose from field experience. Among these were stronger engine and transmission parts, new hydraulic front and rear differential locks, a new common oiling system joining the front differential and final drives to the rest of the cooled and filtered hydraulic oil system, and a new single-lever Quad-Range control.

The new 50 series 4-wheel-drive tractors were announced in the late fall of 1981. Called the "Belt Tighteners," the 8450 and 8650 replaced their 40 series equivalents, with power increased to 185 and 235 PTO and 225 and 290 engine horsepower respectively. Added to these two was the completely new 8850, with a 370-hp V-8 Waterloo engine. It became John Deere's first tractor with 300 hp at the PTO. The new engine, which was turbocharged and intercooled, had been developed over some years for both this and industrial applications.

All models were designed to be adaptable, including use for row-crop work. They all had Quad-Range 16-speed transmissions. The two smaller had Category 3/3N 3-point hitches, the 8850 a 4/4N. Muffler and air stack were relocated to the right side of the Sound-Gard body for better visibility and to allow a new feature, a tilt-up hood for easier servicing.

Another innovation on all three models was the provision of the Investigator II electronic instrument panel with digital readout of engine

The sound room helped engineers at Waterloo develop lower noise levels for the 8440 and other tractors.

rpm, PTO rpm and ground speed. On the left side of the dash there were no less than seven indicator lights and four gauges; a horn augmented the warning lights. HydraCushioned seat suspension and 76.5-78 db(A) sound level in the cab ensured comfort enabling the operator to work long hours.

Innovative Beet Harvester Replaced

The harvesting of sugarbeets offers many unique challenges compared to the more common field crops of corn, soybeans and wheat, which are harvested with a combine. The most obvious difference: Beets are a root crop harvested from the soil, instead of a seed crop standing above ground. The second is the amount of material to be handled; beet yields are typically about 20 tons per acre while corn is only three, with wheat and beans even less.

Advertising for the 4300 beet harvester showed the revolutionary exclusive vertical auger that produced "whistle-clean" beets. The rod grating around the auger further agitated the beets, removing trash, dirt and rocks.

From the '40s on, John Deere sold a variety of beet harvesters that were either tractor-mounted or pull-type units. Each new model was an attempt to improve the beet farmer's income. The sugar mills wanted whole beets with limited damage, no tops and minimum dirt.

The new 4300 beet harvester looked like a real winner when it was first sold in 1973. It appeared to be a much more durable machine because it had eliminated much of the potato chain, a major source of downtime on machines harvesting root crops.

The beet tops were removed as a separate operation prior to harvesting. Conventional digger wheels dug the beets and placed them on a cleaning bed where they were shaken to remove sand, clods, and rocks. The standard cleaning bed was a potato chain with star wheels as an option. Spiral steel grab rolls expelled more rocks and mud as they conveyed the beets from the cleaning bed to the vertical auger. A 4-ton holding tank of proven design permitted continuous harvesting as trucks were being changed.

Advertisements for the 1975 version of the 4300 indicated more than a dozen improvements to handle rocks and mud better and to increase reliability. Even these improvements did not result in the desired functional dependability

in the field, so the 4300 was replaced by the 4310 for the 1977 season. Deere not only had the cost of a major redesign after only four years on the market; they also reworked the beet harvesters in the field to include the new "Ferris wheel."

Why hadn't the experimental 4300 been tested in "all" field conditions before it was produced? It had been tested in multiple states for multiple seasons. The record-setting wet season of 1990 is a reminder that "all" weather conditions are impractical to include. The majority of U.S. corn and soybeans are grown in seven adjacent states stretching from Ohio to Nebraska, while field conditions for harvesting beets are more variable.

Three-fourths of U.S. sugarbeets are grown in California, Minnesota, Idaho, North Dakota, and Michigan. The only two adjacent states are Minnesota and North Dakota, which share the Red River Valley. These five states include small to large beets grown on beds or flatland, irrigated or rainfed, and in an extreme variety of soils with or without rocks. In California, the highest producing state, most users thought the 4300 was great. Users in some other states complained of beet damage, while plugging the auger with mud was a serious problem some seasons in some locations.

There is a limit to how much can be spent for experimental work for a machine like a beet harvester due to the volatility of both the beet market and the beet harvester market. The average price per ton of beets received by the farmer yo-yoed from $16.00 in

John Deere reinvents the wheel, according to the ads for the 4310 beet harvester. The simple wheel elevator minimized scuffing and bruising of the beets while eliminating plugging from rocks or mud.

1972 to $46.80 in 1974 to $21.00 in 1976. These prices do not include Government payments under the Sugar Act but do illustrate the influence of national and international politics. U.S. industry shipments of beet harvesters crashed from 1,278 units in 1976 to only 111 units in 1979. Even at the 1976 peak, beet harvester shipments were only 4% of combine shipments, so experimental costs had to be limited.

The Titan Combines Arrive

Late in 1978 a new series of combines was announced. Stated to be the most productive lineup of John Deere combines ever, the three larger models were equipped with cabs similar in appearance to the Sound-Gard body. They were offered in three versions: grain, corn and bean, and rice and soybean.

The new 6620, SideHill 6620, and 7720 self-propelleds and the 7721 PTO combine offered up to 20% more capacity than previous models, while the new 8820 gave up to 45% more output than any earlier John Deere combine. All had 466-cu.-in. diesel engines governed at 2200 rpm and producing 120 to 200 hp; the concave and beater grate area was increased by 51%, and the straw walkers were 20" longer than those on the models replaced.

Following the success of the Waterloo research and engineering center for the development of

Strengthened up for an increase in power to 225 hp, the 8820 flagship of the combine fleet acquired a new 12-row corn head option in 1981.

tractors, 1979 saw the opening of a combine research center in East Moline, another first in the harvesting world.

All the 1982 Titan combines had the feeder house reverser as standard, and among other improvements were Dial-A-Matic header height control and a pressure gauge indicator for the accumulator. In addition two new platforms announced were the 222 rigid for the 6622 Hillside, 2' wider than any previously used on hillside machines, and a 154" pickup platform that was 22" wider than any previous John Deere pickup, making it ideal for big windrows.

A New Zweibrücken Combine Line

In September 1978 a new top-of-the-line combine, the 985, was introduced in Europe. With six walkers, this was the first John Deere European combine with a 60" wide separator.

In the fall of 1981 the series of combines then current in Europe was replaced by the 10 series. The first models in this new series were the 4-walker 1055 and 1065, 5-walker

The 30-foot grain head was available for the 7720 as well as the Turbo 8820. Harvest capacity was also increased by larger grain tanks and fuel tanks.

Largest of the Zweibrücken-built 10 series combines in 1984 was the 1085 Hydro-4, shown here with SG2 cab, combining a level crop of wheat.

1075, and 6-walker 1085.

All the best and well-proven features of the earlier models were retained—the 24 1/4" cylinder diameter with 104-degree concave wrap, the cross-shaker system for extra grain separation from straw, the 6" straw-walker throw with resulting low 150-rpm speed, and opposed motion of the chaffer and cleaning sieves, adding up to the best threshing combination.

On the Hydro-4 models the engines had some 15% greater horsepower and the grain tanks too were increased in size—in the case of the larger combines, for example, from 4200 to 4800 liters (120 to 137 bushels).

With the introduction of the SG2 cab for the combine line as well as tractors, new controls and instrumentation were adopted to further ease the operator's load, and a new standard of comfort in harvesting was achieved with deluxe seat and air conditioning. The new models maintained John Deere's increasing share of the European combine market.

Another innovation at the same time from the French Arc-les-Gray factory was the introduction of the 1320 mower/conditioner. Destined to set a new trend in mower/conditioners, the 1320 had a 94" rotary cutting mechanism consisting of six oval disks. Its unique conditioning system had 56 V-shaped free-swinging tines that conveyed the grass up and to the rear under the conditioning hood, so that the grass rubbed against other grass, thus conditioning itself. This system was based on research done at the Silsoe research station in Bedfordshire, England. It gave very fast cutting and shorter drying time, together with decreased maintenance costs with its lack of rubber or steel rollers. The new machine was made available to North American as well as European markets.

John Deere Plows a New Furrow

John Deere maintained a reasonable market share in moldboard plows over the years with dependable products, and with several innovations as the market changed. They introduced safety-trip standards on integral moldboard plows in 1950. In 1967, John Deere joined others in the use of hydraulic-reset plows that reset automatically, after an obstruction was cleared, without having to stop the tractor. However, John Deere also used an exclusive dual-pivot standard that permitted a caught share point to move rearward and up as much as 11 inches.

In 1953 John Deere had introduced high-speed light-draft plow bottoms. Tractor power had increased and farmers were no longer satisfied to plow at the speed of horses. But problems remained. Some farmers felt that John Deere plows had higher draft than others (competition in the '60s was primarily from two mainline tractor manufacturers). In the '70s farmers started asking for less plugging with trash. Crop residues had increased and farmers wished to reduce or eliminate any operation to cut the trash before plowing. A shortliner came out with an adjustable-cut plow that offered some time and fuel savings as different field conditions were encountered. Added to all the above was the feeling that the John Deere plow line lacked a

family resemblance and the "Dreyfuss touch."

Thus a clean-sheet approach was taken to plow design just as had been done with the New Generation tractors of 1960 and the New Generation combines of 1970. The first of the new family of plows, the 2600 and 2800 semi-integral and the 3600 drawn, were available in 1979. All three offered a choice of new, more dependable safety-trip standards or dual-pivot spring-reset standards. The 2600 and 3600 improved trash handling with 27-1/2" fore-and-aft clearance and 29-1/2" under-frame clearance. Width of cut (16", 18" or 20") was manually adjustable with shims.

The 2800 featured hydraulic on-the-go adjustment of cutting width from 14 to 24 inches. Fore-and-aft clearance and under-frame clearance were each 33-1/2 inches. Exclusive hydraulic steering of the tail wheel was featured on both the 2600 and 2800 plows.

New NU bottoms were designed for the higher 4- to 5-mph speeds in use at that time. At these speeds, shapes could curve more gradual-

The new semi-integral plows were heralded as the most significant design change in plows since John Deere forged a saw blade into a plow bottom. The 2600, shown with safety-trip standards, had four to six bottoms while the 2800, shown with spring-reset standards, offered four to eight bottoms.

ly and still give adequate inversion and breakup of the soil. In test runs at three test sites, the new plows required 12 to 26% less draft than those of the two main competitors.

The following year, 1980, saw the availability of the 2700 semi-integral plow, a hybrid that offered the manual width-of-cut adjustment of the 2600 with the abundant trash clearance of the 2800 plow. Featured in 1982 was the new 3700 flexible-frame drawn plow with 10 to 16 bottoms. The family line of plows was completed in 1983. The integral 1600 offered the trash clearance of its 2600 and 3600 sisters along with the spring-reset option. The economy 1000 integral plow used the safety trip on its two to five bottoms and offered widths of cut of 12, 14 and 16 inches.

Industry moldboard plow sales suffered a greater decline in the '80s than those of any other major implement. Farmers switched to chisel plows or combination disk-chisels for conservation of time, fuel, water and soil. Time and fuel were saved because a tractor could pull a wider chisel plow than a moldboard plow. In the '70s and '80s, farmers gradually learned that a field that looks trashy after primary tillage makes economic as well as conservation sense.

U.S. shipments of moldboard plows dropped from about 23,400 units in 1981 to only 9,100 units in 1982. At the same time, chisel plows plus combination disk-chisels only dropped from about 17,400 units to 13,100 units, passing moldboard plows for the first time in 1982. Currently, use of moldboard plows is not quite as dead as their sales. Many farmers find it desirable to completely invert the soil every few years for better fertilizer distribution as well as weed and insect control.

Hydra-Push Spreader Based on Customer Invention

Deere receives a steady stream of ideas from outside inventors who suggest new or modified products. Although most of these ideas are not used, each one is examined by one or more people with knowledge in that general area. Frequently the inventor has become frustrated over being unable to buy a product that meets

exact expectations or needs.

While most of Deere's new ideas for products come from their college-trained engineers, educators have been unable to effectively teach innovation. The likelihood of invention is based primarily on an intense feeling of need. The inventor with mechanical ability starts thinking of alternative solutions to the problem. He or she then makes one or more models to determine if the idea really functions as hoped.

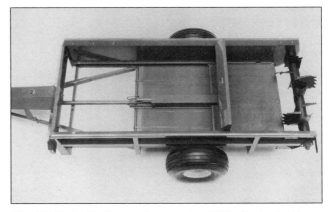

The 450 Hydra-Push spreader, new for 1981, eliminated the apron chain, reduced manual cleanout in cold weather, and provided faster, more uniform unloading. It was joined in 1984 by the larger 780 Hydra-Push spreader.

When the inventor feels reasonably successful, he or she often looks for a manufacturer to use the idea. Inventors whose patented ideas are used are paid a flat fee for the idea, and/or some royalty per machine produced.

The development of the Hydra-Push spreader

Roll-O-Matic front axle was probably the most widely sold customer innovation, as it was a popular option on tricycle tractors for 30 years.

An inventor added tandem wheels to an all-terrain vehicle as the origin of this handy workhorse.

An inventor sold John Deere a working model of a light grader that led to the development of the 570 grader with articulated steering. Deere pioneered this idea, which was copied a few years later by the major builder of graders.

is a representative example of how customer inventions are used by John Deere. Joseph and Daniel W. O'Reilly were dairy farmers near Goodhue, Minnesota. Many dairy farmers in Minnesota and Wisconsin use barn cleaners and haul manure each day of the year as it is produced. Since the acids in manure are very corrosive, the apron chains in the bottom of conventional spreaders eventually rust and break. When this happens, the load of manure must be manually unloaded by fork so the chain can be repaired or replaced. Although the job is considered unpleasant any time of the year, it is devastating when spreading in below-zero weather. Then there is extreme urgency to get the spreader unloaded before the manure becomes one huge chunk of ice that must be removed by a pick.

The idea conceived, tried and manufactured by the O'Reillys was to push the manure to the rear with a movable front endgate or panel similar to that used in unloading most garbage trucks. To get it to work dependably under a variety of conditions they found it necessary to use plastic surfaces on the floor and sides of the spreader. They also found they needed a movable floor in the front of the spreader to start the load moving to the rear.

With the functional performance of these three related ideas well proven in the field, Welland Works engineers proceeded to develop two sizes of spreaders based on these principles. Numerous changes were made relative to the spreaders while keeping the basic principles. The plastic liners, which had been riveted to the sides and floor, were replaced with fiberglass-reinforced plastic bonded to the sides and floor. The very long push-off cylinder for the combination panel and movable floor was replaced with tandem cylinders. The lower cylinder pushed the panel and floor together for the first two-thirds of the load. Then the second cylinder completed unloading by pushing the panel only. This design change permitted the tongue to be much shorter and avoided extending the movable floor out the rear of the spreader.

Several other John Deere machines or components have had input from outside inventors in their development. These include the combine straw spreader, the hillside combine, and the Tru-Vee opener and finger pickup seed meter on Max-Emerge planters.

The bale ejector introduced in 1958 made haying a one-man operation by throwing the bales into the wagon, eliminating the need for a second person to stack bales.

Patents Provide Protection Sometimes

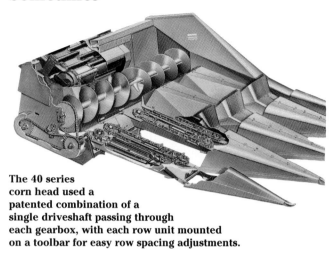

The 40 series corn head used a patented combination of a single driveshaft passing through each gearbox, with each row unit mounted on a toolbar for easy row spacing adjustments.

National patent systems are designed to promote innovation. In the U.S., a patent gives the owner the legal right to exclude others from making, using or selling the invention for a period of 17 years. For this exclusive right, the inventor must publicly disclose a clear description of the invention and how to best make and use the invention. This provides other inventors with an opportunity to consider alternative approaches to a better solution to the problem, so long as they do not infringe the claims allowed in the patent.

Most farm equipment has patents dating back several decades before that machine became popular on the farm. That is especially true of combines and corn pickers. The bale ejector, a Deere product development department project in the mid '50s, was an exception to that case as there was no prior "art" (history). When the patent on the ejector was granted, a company attorney reported gleefully to Murray Forth, the engineer who developed it, "Anyone who tries to copy our ejector will have to dribble the bale before shooting it into the wagon!"

However, Deere's two major baler competitors were soon selling bale throwers that neither infringed the patent nor "dribbled the bale." How could this be? Deere & Company's 20 patent claims repeatedly discussed the "throw-

ing stroke." The competitive throwers used no throwing stroke, as they ran continuously. One squeezed the emerging bale between belts and the other used rollers. The competitive throwers did not look like the John Deere unit but achieved similar results.

The patent protection of the 40 series corn head, a much more important product for Deere and for competition, was effective. The corn head is more a marriage of several good engineering designs than any single spectacular innovation. Besides functioning very well in the field, it required fewer adjustments and less frequent lubrication of fewer lubrication points.

QC Times 10/12/82

Decision socks IH

By David M. Schechter
of the Times

International Harvester Co. has been ordered to pay Deere & Co. nearly $28.5 million for infringing on a Deere corn head patent.

The damage award was entered Friday by U.S. District Court Judge Robert Morgan in Peoria, Ill.

Harvester plans to appeal the damage award, spokesman Bill Colwell said.

"We have been advised by John Cline, of the law firm of Hume, Clement, Brinks, William & Olds Ltd., that 'in his judgment, the decision is clearly erroneous,'" Colwell said.

The appeal will be filed with the Court of Appeals for the federal circuit in Washington, D.C., which specializes in patent matters.

Deere spokesmen could not be reached for comment this morning.

The exact damage award — covering Harvester's use of the device from 1974 to 1980 — was $28,462,664, a court spokeswoman said.

A corn head is a large piece of

equipment on the front of a combine that strips corn from the stalk.

Morgan ruled in U.S. District Court in 1978 and again in 1980 that Harvester infringed on the patent. The second decision came after Harvester won an appeal of the case.

The U.S. Supreme Court last November refused Harvester's appeal of a ruling in Deere's favor handed down in July by a three-judge appeals court.

The original suit was filed in 1976 in U.S. District Court, Rock Island. Morgan retired from active service on the bench in May and now holds senior status on the court.

The patent suit applies only to the 800 series, which Harvester ceased manufacturing in 1980.

In a hearing in July, Keith Williams, Deere's manager of cost accounting services, testified that Deere's share of the national farm implement market grew steadily to 42.2 percent in 1973 — the year before Harvester introduced its corn head.

From 1974 to 1979, he said Deere's share of the market dropped.

A protracted patent suit led to a large settlement in Deere & Company's favor. It was many times greater than the company has ever had to pay for a patent.

1982–1988 A Period of Consolidation

During the farm and industrial downturn that brought mergers and realignment of its competitors, the company tightened its belt and took advantage of the new facilities. The emphasis was on value and the maintenance of existing equipment, pending the eventual improvement in the market.

Caster/Action Mechanical Front-Wheel Drive

Despite the economic climate, the fall of 1982 saw the most extensive list of new tractor models announced by the company. The five large row-crop tractors from Waterloo and the five medium tractors from Mannheim in the new 50 series offered the option of Caster/Action mechanical front-wheel drive. This exclusive patented design with 13-degree caster tilted the front wheels to permit a much shorter turning radius.

John Deere had pioneered hydrostatic power front-wheel drive on Waterloo tractors in 1968 and power rear-wheel drive on combines in 1977. Hydrostatic drive for tractors had been chosen because it offered a short turning radius, easy wheel tread adjustment, and adequate

crop clearance with reasonable-size front wheels. However, it did not add enough drawbar pull to be a popular option. Caster/Action mechanical front-wheel drive (MFWD) overcame the turning problem and provided up to 20 percent more drawbar pull.

Hydrostatic rear-wheel drive is still a good option on combines because it provides more torque than the tractor units did. Also, combines have no drawbar loads but are simply trying to get through soft, muddy fields.

Deere and other U.S. tractor manufacturers lagged behind European manufacturers in offering differential lock and mechanical front-wheel drive. European farmers contend with wetter fields, and readily adopted these two features to increase traction.

The new row-crop models were the 100-hp 4050, 120-hp 4250, 140-hp 4450, 165-hp 4650, and a new top-of-the-line 190-hp 4850. The 4250 was also available in a Hi-Crop model. The tractors featured a number of innovations including a new 15-speed Power Shift transmission, optional on the four smaller models and standard on the 4850. It provided effortless on-the-go shifting, with seven well-spaced field speeds, four speeds below 3 mph for PTO work, and four speeds for transport. The 16-speed Quad-Range transmission was standard on the 4050 through 4650.

With improved power, fuel economy and hydraulic efficiency, these new tractors represented another significant step forward. Each model had 10 hp more than the tractor it replaced. State-of-the-art engineering reduced fuel consumption with John Deere's exclusive viscous fan drive, reduced-friction pistons, and longer-stroke fuel pumps. Hitch sensing was hydraulically assisted.

Caster/Action MFWD provides the short turning radius needed for row-crop farming and for loader operation.

Mechanical front-wheel drive gave the 4650 and 4850 tractors the traction for heavy loads that had previously been available only on 4-wheel-drive tractors.

The new Sound-Gard body was quieter and incorporated an Investigator II warning system which gave audible and visual warning of malfunction in any of the tractor's systems.

Kansas City was the place for the 1985 new product announcement. Emphasis during the continuing difficult U.S. farm market was on factory efficiency and dealer support. The company also emphasized safety on farms by introducing a "Get your Roll-Gard up" program. Rollover protective structures (ROPS) and seat belts were standard on all John Deere farm tractors manufactured after November 1, 1984, except the 2255

The 50 series 4-wheel-drive tractors were popular for air seeding and other operations in Western wheat areas.

orchard/vineyard tractor. Special reduced prices in 1985 were offered for retrofitting this equipment on most tractor models back to 1960.

Yanmar Joins Farm Tractor Line

To match the new 100- to 190-hp tractors, the medium-size 45- to 85-hp models also became part of the 50 series. Advertised as "Efficiency Experts," the five models announced were the 45-hp 2150, 55-hp 2350, 65-hp 2550, 75-hp 2750 and 85-hp 2950. Each offered 5 more horsepower than its predecessor.

The 2150 had a 3-cylinder engine; the three middle models were 4-cylinder, the 2750 being turbocharged, and the 2950 was 6-cylinder. All were available with mechanical front-wheel drive. Caster/Action drive with 12-degree tilt was used on all models except the 3-cylinder 2150.

The Sound-Gard body was a new option on the 2750 and other 4-cylinder models. It was not offered on the 3-cylinder 2150 because of interference with loaders, top-heavy appearance, and anticipated low sales.

Two additional models, the 3-cylinder 50-hp 2255 orchard and 2255 vineyard, were similar in most ways to the 2150 but were built lower for specialty work. They were distinguishable by their rear wheels, 18.4-16.1 in the case of the orchard model, giving it a 65.6" overall width, and 14.9-24 for the vineyard with a width of only 59". Another similar tractor was announced, the low-profile 2750 for orchard and grove.

The 2750 Mudder had mechanical front-wheel drive as standard, with large-diameter rice tires. This gave vegetable growers the crop clearance needed for cultivation and the traction required when harvesting in muddy fields.

With the declining farm economy of the early '80s, John Deere did not want to abandon the 40-hp slot so it added the Yanmar-built 1250 in 1982 to fill the vacancy. The 1984 season saw the final extension of the Yanmar line of tractors with the introduction of the 50-hp 1450 and 60-hp 1650. These tractors provided the customer with a lower-cost alternative to the Mannheim-built tractors, which had 5 more horsepower and considerably more weight.

Standard specifications included power steering, wet-disk brakes, nine forward and three

Mechanical front-wheel drive was a useful option for the 1650 and other Yanmar-sourced tractors for loader work.

reverse speeds, 3-point hitch, and easy servicing with flip-up hood.

The 1650 broke the diesel tractor fuel economy record at Nebraska, held until then by the 2-cylinder 720, with a final figure of 18.81 horsepower-hours per gallon. This extreme efficiency was achieved with optimum design of the combustion chamber and fuel delivery system.

The new 3150 was advertised as the replacement for that most popular of tractors, the 4020. It had the same horsepower and about the same dimensions, but double the number of speeds, better fuel economy with greater torque reserve, and mechanical front-wheel drive as standard.

The 900HC tractor with 23" ground clearance for tobacco and vegetable growers was introduced in 1986. This Yanmar-built tractor with off-set driving position was reminiscent of the Models "L" and "LA" of the '40s. It was available with a creeper 0.77-mph first gear, mid-mounted cultivator, and Category 1 3-point hitch.

Another attempt to match the farmer's reduced income came in 1986 with the addition of the 55- to 85-hp "Price Fighter" tractors. This alternative line offered 5 to 9 percent savings by using flanged rear axles with steel wheels and other lower-cost options.

John Deere celebrated 1987, its 150th anniversary, by introducing the new 55 series tractors in the 45- to 85-hp class. These efficient tractors cut fuel consumption up to 10% in hard conditions, more in light. A TSS (top-shaft-synchronized) transmission was standard on the 4-cylinder 2755 and 6-cylinder 2955, and optional on the

The 2155 was the only model without tilt steering wheel, hydrostatic power steering, and the options of Sound-Gard body, 1000-rpm PTO, Caster/Action MFWD, and rack-and-pinion axles.

3-cylinder 2155 and 4-cylinder 2355 and 2555.

Although 29 new machines had been added to the line in 1987, it was largely a question of consolidation pending the recovery of the farming market from its mid-'80s doldrums.

The 55 series tractor line was completed in 1988 by the replacement of the 3150 with the 3155, which had a TSS 16-speed transmission with hydraulic Hi-Lo.

Mannheim Celebrates 125 Years

For the orchard farmer and vine grower in Europe, five new compact tractors were introduced for the 1983 season. The 3-cylinder 1040V and 1140V vineyard models had a new smart hood style, a creeper gearbox option, and new high-power closed-center hydraulics. More legroom and a suspension-mounted cushion seat, plus a folding rollbar, gave the operator more comfort and safety.

Headlights were built into the hood to give added clearance. The three orchard models, the 3-cylinder 1140F and 4-cylinder 1640F and 2040F, had an additional Hi-Lo transmission option. All five had hydraulic wet-disk brakes and could have power steering.

A new line of "economy" tractors was introduced combining quality features with an economy price, and was well received in Europe. The X-E series was offered in the 1640, 2040, 2040S and 2140 4-cylinder versions and had the SG2 cab, optional Caster/Action MFWD, and synchronized transmission with Hi-Lo option.

At the Mannheim Works, formerly the Lanz factory, a century and a quarter of operation was noted by building the 750,000th tractor in 1984. Mannheim continued its record of the previous 11 years as the largest manufacturer and exporter of farm tractors in West Germany.

The 4-post Roll-Gard structure is mounted on the fenders of the 4-cylinder 2140. Mannheim 3- and 4-cylinder tractors sold in the U.S. used 2-post Roll-Gard structures mounted to the axle. Also note the front fenders.

The big news abroad during 1984 was the announcement of the new 112-engine-horsepower 3640 tractor with the exciting new multi-implement capacity of front hitch and front PTO. Similar in overall layout to the 3040 and 3140, with its Caster/Action mechanical front-wheel drive, it featured virtually 50-50 front-to-rear weight ratio and was designed to handle big loads with maximum efficiency and stability.

The European line now consisted of 13 models (all power ratings given here as PTO hp)—the 840 (34 hp), 940 (39 hp), 1040 (44 hp), 1140 and 1140N (49 hp), 1640 (54 hp), 2040 (61 hp), 2040S (66 hp), 2140 (72 hp), 3040 (78 hp), 3140 (84 hp), 3640 (98 hp), 4040S (98 hp), and 4240S (115 hp).

To fill the gap between the 132-engine-hp 4240S and the 160-hp Waterloo 4450, a new model specially designed for Europe was announced for the 1986 season. This was the 4350 with MFWD and Category 2-3N 3-point hitch.

Also introduced in Europe, for the stock farmer, was a new economy MC1 cab with a low profile for 3- and 4-cylinder tractors from the 1040 through 2040S. The 1040 was also offered as a highway tractor painted industrial yellow and black.

In 1986 Deere sales volume and market penetration increased in most of the larger European markets, including France, Spain and Germany. So, the announcement of a completely new tractor line for 1987 was timely. The chief improvement of the 50 series was better fuel economy. Contributing to this was the new self-regulating Eco-Fan, which saved up to 0.8 liters of fuel per hour by automatically adjusting its speed to match actual cooling needs. High piston-ring position, low-friction pistons and reduced engine speeds also helped achieve economy.

The Mannheim-assembled 4350 included European front lights, and hitch pin in the front weights.

The 50 series consisted of the 3-cylinder 1350, 1550, 1750 and 1850, the 4-cylinder 2250, 2450, 2650 and 2850, and the 6-cylinder 3050, 3350 and 3650. The 4350 was continued, plus the 4450 through 8850 from Waterloo. Innovations included a 1-mph first gear when equipped with the Power Synchron transmission, and an optional 25-mph high-speed version on the tractors over 60 hp.

A new turbocharged version of the 3-cylinder 1850, the 1950, was introduced for the 1987 season, giving about the same power as the smallest

The 2650 is shown with the optional front hitch and PTO, available on all seven 4- and 6-cylinder Mannheim tractors. European farmers use many front and rear combinations of equipment, partially because of transport width limitations.

of the 4-cylinder models, the 2250. This gave the company a very comprehensive range of five 3-cylinder, four 4-cylinder and three 6-cylinder models from their Mannheim factory.

To these models were added the 45 series orchard tractors built in Italy by Goldoni with unique styling, though still very recognizable as part of the John Deere family. The four models were the 42-engine-horsepower 1445F, 48-hp 1745F, 56-hp 1845F and 67-hp 2345F, all with three cylinders, the largest being turbocharged.

Specialized versions of the Mannheim models were built in Getafe, Spain, including four F (fruiteros), two M (multicrop or high-crop), and two V (vineyard) models, the relevant letter following the model number on the decal. S was for standard 2-wheel drive and S-DT (standard double traction) for MFWD models.

Overseas Sales and Profits Increase

In Brazil the company's affiliate had its best year ever during 1983-84. In 1985, increased combine capacity was planned. John Deere received an order from China for $25 million worth of equipment including 400 large tractors, 500 drills, 200 cultivators and 100 European combines. Markets were particularly strong in the United Kingdom, Spain and Mexico during 1985.

To celebrate 30 years in the Mexican market two new models were announced in 1984, the 2755 and 4255. The company also agreed to supply components to the local Mexican government operation, Sidena, to build a 60-hp tractor. The former International Harvester factory at Saltillo was purchased and renovated for the production of the 2755, 4255 and 4455 tractors.

The 2755 Turbo was one of the models sharing in the large gains in the Mexican market from increased production in 1986.

A Period of Consolidation

The Sedan line of tractors in Australia was replaced in 1982. The new models were the 94-PTO-hp 4090, 110-hp 4290, 129-hp 4490 and 154-hp 4690. The two smaller models had 359-cu.-in. engines, while the two larger had 466 cu.in. displacement. For the first time in Australia, the new series were painted John Deere green and yellow and had the company's familiar styling, but still retained the Chamberlain name and emblem—a television screen-shaped ellipse with an outline map of the country and the word Chamberlain across the center. The new tractors' parentage was nevertheless very obvious.

The company's ownership of the Chamberlain organization moved from a minority to a strong majority position during 1986. That same year, a decision was reached to give up making special tractors for the Australian market in Welshpool.

In South Africa, the 41 series tractors proved so reliable and gave such excellent performance that farmers accepted this new concept of basic Mannheim design with local engines. As a result, in September 1987 the new 51 series tractors were announced, with 16 models in six sizes, from the 2251 to the 3651. The 2651 TransTill model was capable of 25 mph on the road, for use mainly by sugarcane and timber growers.

The 50 series tractor models which were announced from Rosario, Argentina, for the 1988 season were the 4-cylinder 2850 and the 6-cylinder 3350 and 3550, the latter with MFWD as standard. On the 3350 it was as an option. They had 95, 110 and 125 engine horsepower respectively.

Deere & Company Leadership Excels

Deere & Company net income had peaked in the good farm economy year of 1979 at $311 million. Net sales to dealers continued to rise and crested in 1980 at $5.5 billion. In October 1982, when William A. Hewitt retired and Robert A. Hanson took over as chairman, no one knew the length or depth of the farm depression ahead, or the devastating effects it would have on the farm equipment industry as a whole. The bottom was reached in 1986 when the company's sales were

Robert A. Hanson's skill, courage, enthusiasm and warmth helped management stay the course during the turbulent '80s. The remainder of the North American farm equipment industry experienced major traumatic restructuring during this period.

$3.5 billion and it reported a net loss of $229 million due to market conditions and a major strike. While the company also lost money in 1987, its 150th anniversary year, conditions improved significantly with the end of the strike and a modest increase in sales.

Throughout this period Deere carried on a broad program to reduce costs and improve efficiency. Factory breakeven points were cut substantially as worldwide employment decreased from 65,400 to 37,800. Research and development efforts and capital improvements continued at significant levels. During the restructuring, no North American city lost its John Deere factory. Moline did see the merging of its Plow & Planter Works with the East Moline Harvester Works, and the transfer of tillage products to the Des Moines Works.

Boyd C. Bartlett served ably as president from February 1985 until his retirement in March 1987 after 35 years of service. Hans W. Becherer then became president after heading the company's worldwide agricultural and consumer products operations. Before Hanson's retirement as chairman in May 1990, he saw both sales and net income set new records in 1988 and 1989. His leadership team had succeeded!

Titan II Combines Announced

The Titan combine line had the addition of HarvesTrak monitors to help keep losses low and harvest capacity high. A new widespread straw

The 7720 and other Titan II combines featured chrome-plated and hardened rasp bars for soybeans and other abrasive crops. Dual wheels were available for better mobility in muddy conditions.

chopper spread the straw up to 30 feet. Very wide 73x44-32 flotation tires were available for the 7720-R rice model and all 8820s. Added were new Row-Crop heads, the 854A for eight wide rows and the 1253A for 12 narrow rows, with a sump-type 24"- diameter cross auger giving steady delivery of the crop to the feeder house.

The Titan II combines came out in 1985 with a 7" longer shoe, giving up to 15% more capacity in shoe-limiting conditions. Lift-out sections at each rear corner of the shoe reduced grain loss on slopes.

To give those farming in hillside areas greater harvesting capacity the new 5-walker 7722 Hillside model was introduced in 1988. An innovation in the combine world was the introduction of the new 900 series platforms with stainless steel transition sheets, a John Deere exclusive for smoother flow from the cutterbar to the auger. Flexible platforms up to 24' and rigid up to 30' were available.

Unique Combine in Australia

Wheat yields are often limited by moisture availability. In Deere's main combine markets the highest yields are in Europe, followed by Mexico, the U.S., Canada and Australia. Australian low yields require wide headers and minimum investment. A special tractor-drawn model was built for the Australian market. It was PTO driven, and had a revolutionary 25'-wide platform that was permanently mounted on a unique 2-wheel cart. This allowed the use of a much wider platform than would otherwise have been the case, and in addition the machine could be converted from transport to working position by one person in less than 5 minutes. The design also made a 225-bushel grain tank possible.

The controls for the platform drive and height, unloading auger, swing-out, cylinder speed, separator drive and platform cart swing cylinder

A Period of Consolidation

Parclem CJD

The 1051 PTO combine platform, designed for short-straw wheat, had a closed front, twin augers and no reel.

were all located in the tractor cab. In-cab monitoring included grain loss, grain tank level, cylinder speed, straw walker blockage and low shaft speed of walkers, fan and both elevators. The threshing unit had the standard Zweibrücken 24 1/4"-diameter rasp bar cylinder, the largest used by major manufacturers up to that time, and four straw walkers with 6" throw on the crank, another secret of John Deere combine success.

The first model of the new 1100 series combine line was introduced in Europe for the 1987 season in two versions, the 1188 and the 1188 Hydro-4. An important contribution to grain saving was the ability to control the fan speed from the cab. A lever beside the operator's seat gave instant stopping of the header and feeder house, with a hydraulic reverser to eject plugs

or foreign bodies. The new models featured easy cleanout, for quick switching between different crops, and 16-mph road speed.

The new 800 series cutting platforms, with extendable table, Dura-Cut cutterbar, full-width skid shoes—very necessary for cutting low to collect all the straw as required in European conditions—and a gearbox knife drive, were the most adaptable on that market. Also offered were headers for sunflower, corn, and rice, plus a flexible soybean platform and a quick-mounting rape attachment.

Following the successful introduction of the first model of the 1100 series in 1987, further models were added in 1988—the 180-hp 1177 Hydro-4, the 150-hp 1177, the 125-hp 1174 and the 125-hp 1166, the latter with four straw walkers.

The 1188 Hydro-4 combine featured the best cab in the industry. The SG-2 provided in-cab tailings check (the best visual guide to the machine's performance), grain tank sampling, and 80-dB(A) sound level. It had a special monitoring system and a master control lever that gave fingertip command of ground speed and header.

1988–1990 Maximizer Combines Ready for the '90s

Palm Springs, in January 1989, was the place for the largest-ever new product announcement by Deere. A completely new combine line was augmented by the Waterloo 55 series tractors, a new line of round balers, the 600 series disks, and the market's first 5-row cotton picker.

60 Series 4-Wheel-Drives Answer Farmer Requests

The 60 series 4-wheel-drive tractors, from the 198-PTO-hp 8560 to the 256-PTO-hp 8760 to the 322-PTO-hp 8960, had been introduced in October 1988 in Denver. Over the years, owners of John Deere 4-wheel-drive tractors had asked for various improvements. These new tractors were designed from the ground up to meet these requests. Daily service time was reduced to 2 minutes. The modular-designed chassis greatly reduced the time required for major overhauls of engine or transmission.

The 8960 had a 370-hp, 14.0-liter, 6-cylinder Cummins engine that was turbocharged and water-to-air aftercooled. The previously used John Deere V-8 had been a good engine but production quantities were never high enough to obtain competitive costs. The other two tractors continued with upgraded John Deere 7.6-L and 10.1-L engines, which were air-to-air aftercooled.

All three could be equipped with a 12-speed Synchro or a 24-speed PowrSync transmission, while the two larger models could also have a 12-speed Power Shift. All could have a 1000-rpm PTO option, regardless of transmission, for factory or field installation.

An artist traces the ancestry of the new 8760 tractor back through the 8430, the 7520 and the 8010.

The 8960 is especially liked in the broad expanses of the Western wheat country.

These new models were all the same in overall dimensions with a 134" wheelbase, longer than those they replaced, but each had its own detail differences. They all had a turning radius of 14'6" with their 42-degree turn angle.

The largest model had hydraulically power-actuated wet-disk brakes on both axles; the two smaller models had brakes on the front axle only. The new styling emphasized that these were completely new tractors from the ground up.

Improved Engines in Large 55 Series Tractors

More than 6,000 dealers and press representatives visited Palm Springs, California, in January 1989 to see the largest assembly of new agricultural machines announced at one time by the company. Dealer personnel from 1,800 North American agricultural dealerships attended during the 3-week event. Thirty countries outside of North America were represented by 400 people.

Redesigned 7.6-liter (466 cu. in.) tractor engines provided easier starting, more lugging ability, cleaner fuel burn and greater fuel economy. In fact, the 4955 set a new fuel economy record in the Nebraska tests. Combustion effi-

ciency was improved by better air flow from larger turbochargers and improved ports and valves. Combustion was also improved by new pistons, fuel injection system and nozzles.

The 4555 is a new size, filling the gap between the 140-hp 4455 and the 175-hp 4755.

The new row-crop series introduced a number of innovations to this size of tractor, including Deere's first 200-hp model. The three smaller tractors retained the 50 series power ratings; the new model at 155 hp was the smallest of the three larger models. Power for the others was increased by 10 hp, to 175 hp for the 4755 and

The 200-horsepower 4955 has sufficient power and traction for plowing a wide swath in heavy alfalfa sod.

200 hp for the 4955. The three smaller models had 65-gallon fuel tanks; the three larger, 102-gallon. The 4555 is the only model not exported to Europe.

The mechanical front-wheel drive was automatic—when set in "auto" mode it disengaged as either brake was applied for turning, and it reengaged when the brake was released. On the road, the front drive disengaged above 9 mph to save tire wear. Applying both brakes provided instant 4-wheel braking for safer stops.

The new Sound-Gard body had improved panoramic view, HydraCushioned swivel seat and a more comfortable interior. Added to this was a new IntelliTrak monitor system for watching all important tractor functions from routine service maintenance to ground speed, distance traveled, area worked and area covered per hour. An optional radar sensor gave true ground speed and percent of wheel slip.

Even the wheels were redesigned with better mounting in place of the former rim clamp arrangement, eliminating tire wobble. Finally, in new innovations the 3-point hitch on the 4555, 4755 and 4955 had electronic sensing, giving a blend of load control and depth positioning, plus control of hitch lowering speed and raise height.

The 4255 was available in a Hi-Crop model for growers with crops on tall beds. In addition to the 75-hp 2755 High Clearance model (which replaced the 2750 Mudder), a new 6-cylinder 85-hp 2955 High Clearance tractor joined the vegetable farmer's line. Also of use to him and to tobacco farmers, Wide-Track versions of the 2355, 2555, 2755 and 2955 2-wheel-drive models replaced the similar 50 series, with 96" front axles with 25" clearance.

In the orchard and vineyard the 50-hp 2255 was replaced with the 55-hp 2355N, and an additional model, the 80-hp 2855N, was added to the line. Both models were equipped with telescoping Roll-Gard ROPS as standard equipment, and they both stood less than 59" tall.

The 4255 is a nimble tractor for the livestock farmer, with the ability to carry heavy drawbar loads.

The 2355N tractor remains just under 65 inches wide, even with the optional Caster/Action mechanical front-wheel drive.

John Deere Builds Three Millionth Diesel Engine

On October 12, 1990, at its Engine Works in Waterloo, Iowa, John Deere celebrated manufacturing its three millionth diesel engine. This record includes engines produced at the Waterloo and Dubuque, Iowa, factories along with those made in Saran, France. The Saran factory built its 900,000th engine on October 17, 1990.

John Deere is the leading producer of 50- to 300-hp diesel engines for the off-highway market in North America.

Deere concentrated on all-fuel engines for the first thirty years of building tractors. These engines were started on gasoline but switched to run on lower-cost kerosene or distillate. Gasoline engines rapidly replaced the all-fuel option in the '50s because they had more power.

Liquified petroleum gas (LP-gas) engines were available in the '50s and '60s. LP-gas offered the same power as gasoline, cost less to buy in areas where it was produced, and reduced the frequency of engine overhauls, but was difficult to handle.

Deere's first diesel engine was introduced in 1949 on the Model "R" with 43 hp. Diesel engines captured over half of the engine production at Waterloo as soon as the New Generation of Power was introduced in 1960. Gasoline engines are easy to start and cost less than diesels, so remained the preferred engine for tractors below 50 hp in North America until 1967. However, the greater durability and lower fuel cost of the diesel won out, so Deere switched entirely to diesel engines for farm tractors in 1973.

Waterloo builds more than 30,000 engines each year. Its 6-cylinder 466-cu.-in. engine powers tractors from 105 to 200 PTO hp. An electronic fuel control is used on this engine for the 9500 and 9600 combines. On some 9600s, this electronic control boosts power to 253 hp for unloading on the go. Waterloo also builds the 6-cylinder 619-cu.-in. engine used in the 8760 tractor at 256 PTO hp.

Dubuque and Saran produce the 300 series engines in 3-, 4- and 6-cylinder models. Displacement varies from 179 cu. in. to 414 cu. in. They are offered in naturally aspirated, turbocharged, and turbocharged with aftercooling versions.

Deere produced more than 117,000 engines in its three factories in 1990. About 75% of these were used in John Deere farm tractors, combines and industrial equipment. The remaining 25% were sold by John Deere Power Systems to "original equipment" manufacturers for air compressors, electric generators, irrigation pumps, boats, etc.

Combines Pamper Operator and Give Superior Cleaning

Two major decisions were made seven years before the birth of the completely new 9000 series combines. Conventional straw walkers were to be used and the cab was to be centered, with the engine back of the grain tank like the 55 and similar combines.

Combines are the most complex machine on farms, with many functions that must be controlled. Farmers also work quite long hours during the harvest season. Thus, great emphasis was placed on developing the cab, controls and mon-

The cab offers a panoramic view, convenient controls, comfort and the popular buddy seat.

itors to make the operator's task easier. The well-sealed cab is dust free and has low noise levels with the engine behind the grain tank. The centered cab offers a wide angle of vision, while the longer feeder house makes it easier to observe the header. The controls most often used, header height and reel lift, are located on the hydrostatic ground speed control handle. Other switches are on the right console and are both color and shape coded. Monitors for a wide variety of engine and combine functions are found in the right corner post and overhead. The "buddy seat" has found a variety of uses including training new operators.

Operator safety was considered throughout the design of the combine. Essentially all functional controls except sieve adjustments are in the cab. All action-function switches must be in the off position for the engine to be started. A reverser on the feeder house backs out plugs. A seat switch shuts down the header if the operator leaves the seat. Some of the other safety features seem more indirect but are also important. Belts and chains were reduced by 33%, resulting in less need for adjustment and replacement. The big side shields of the combine swing up out of the way for easy service. Both the number of lubrication points and the frequency of lubrication were reduced, making servicing 50% easier than on previous models.

At about $100,000, the most expensive machine on the farm should have a good appearance. Dreyfuss styling on the new combines includes very few visible spot welds on the sheet metal. The entire separator body of the combine is dipped in an E-coat paint system for protection of the inside as well as the outside. A finish coat of polyurethane provides a rust-free shiny coat for many years.

The highest capacity combine John Deere has ever made, the 9600, has a 64.5" wide cylinder, five walkers and a 240-bushel grain tank. It is powered by a 466-cu.-in. turbocharged engine providing 200 hp in the grain models and 253 hp in the corn special. The 9500 is quite similar but narrowed to four walkers, with a 53.5"-wide cylinder and a 204-bushel grain tank. It uses the 466-cu.-in. turbocharged engine at 200 hp. The

The 9600 combine with 1243 corn head fills the 240-bu. grain tank in less than 10 minutes in typical corn yields.

The biggest functional innovation in these combines is the Quadra-Flo cleaning system. Four fans on a common shaft provide a uniform flow of air across the width of the combine. Most of the air flows to the rear to a conventional shoe. Some of the air flows forward to the precleaner, which suspends much of the chaff before it has a chance to become mingled with the grain. Some of the grain goes through the precleaner and on directly to the grain tank. Operators find this cleaning system requires fewer daily and field-to-field adjustments. Fan speed is adjusted from the seat.

Factory workers had more input on how these combines should be built than in the past. They were consulted in the selection of factory tooling and its layout. The Harvester Works production areas were completely reworked for the new models to get a better product at a reasonable cost.

Customers were consulted as to their needs before, during and after the development of the combines. They put some of the 36,000 test hours on the experimental combines. Market research surveyed 256 owners of the new combines in 1989. More than half of all new own-

9400 keeps the cylinder width and number of walkers of the 9500. However, the walkers are shorter, the grain tank holds 182 bushels, and the 359-cu.-in. turbocharged engine is used at 167 hp.

Functional performance was improved in numerous ways. Threshing received a new 10-rasp-bar cylinder 26" in diameter or 18% larger than previous John Deere combines. The 14-bar concave wraps farther around to provide better threshing and separation. The combination of the larger cylinder and larger-area concave permits slowing the cylinder to reduce grain damage. The majority of grain is separated by the cylinder, concave and beater. Also, the new walkers were designed with more-aggressive front four steps and a greater length.

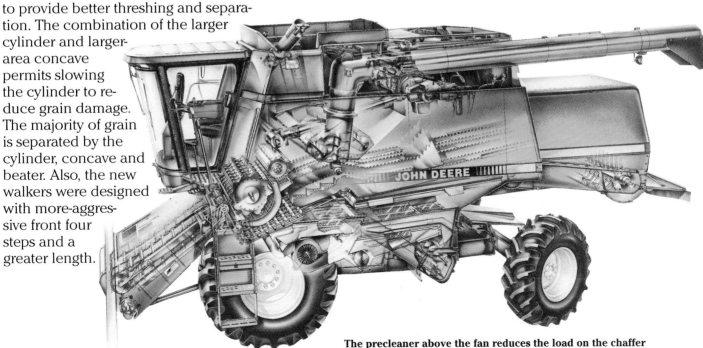

The precleaner above the fan reduces the load on the chaffer and sieve for a cleaner sample in the grain tank.

ers were contacted in the field or on the phone by Harvester engineers, management, or production workers. Customer acceptance of the new Maximizer line assures Deere's combine sales in the '90s, just as customer acceptance of the New Generation of Power assured tractor sales in the '60s.

Pregnancy Still Lasts Nine Months

A popular topic at engineering meetings for at least the past forty years has been some new technology that should greatly reduce the total time required to develop a new product. But this expectation has proved about as realistic as to expect a skilled female obstetrician to have a shorter than normal pregnancy.

Many of the technologies that seemed dramatic when introduced have become routine tools used by engineers and management. Examples include stress analysis, reliability forecasting, computer-recorded and analyzed field test data, telemetry for collecting field test data, cellular manufacturing, certified suppliers, just-in-time supply, computer-aided design, computer-aided manufacturing, and simultaneous engineering.

Counterbalancing these new technologies are the many new demands on the developmental process. Customers drive the whole system, so their demands come first. They want equipment that will function well in a greater variety of crops, yields, soils and moisture conditions. They want harvesting equipment to work more hours of the day with high capacity and low losses. They want to minimize lost field time from malfunctions or parts failures. They want to reduce the non-productive time spent in routine lubrication and maintenance as well as the time spent in going from transport to field operation. They want safety but do not want protective devices to interfere with getting the job done. They want equipment to be "user friendly." Customers are in a competitive business so they expect competitive prices for the equipment they buy.

The dealers want equipment that is convenient to unload when received and simple to set up for customers. It should be easy to demonstrate to prospects and owners. Equipment failures should be simple to diagnose and correct or repair.

Marketing people want equipment with demonstrable features and productive advantages over competition. They want as many models and options as appear to be needed to meet a wide variety of customer requirements. They want minimum lead time for development with a reliable end product. They also want a competitive advantage on pricing.

Manufacturing people want parts that can either be made on standard tooling or that require minimum changeover time between different parts. They want a minimum number of models and options, to hold their costs down. They want maximum use of standard parts and a limited number of suppliers of purchased parts.

Society has added its requirements to the developmental process also. The health of the customer, the factory worker and the community as a whole must be protected. Noise levels of both the product and the manufacturing process must be controlled. Servicing the refrigerant in air conditioners should not pollute the air. Painting the equipment should be done in a way that protects both the air and water discharged from the factory.

One favorable outcome of the competitive squeeze of the '80s in America was a reintegration of product engineering with manufacturing engineering. These two groups had become more specialized, drifting farther apart in their work and tending to work in sequence rather than together. Most John Deere factories now have a single manager of engineering over these two functions. The result is saving of developmental time, and more easily manufactured parts that perform well in the field.

However, there remain many things that must be done in sequence in the developmental process. Customer requirements are gathered by factory marketing, management and engineers along with similar inputs from Deere market research and product planning. After the specifications for a new machine are agreed upon in writing, a backlog of functional improvements goes into an experimental machine. Customer safety, comfort and convenience are integral parts of the entire developmental process. Major new designs require multiple years of testing in a variety of field conditions. As more is learned about function, modified parts are installed or complete new experimental machines are built. Reliability testing of components in the laboratory has made great strides, but functional testing still relies primarily on field testing. This must include use by potential customers as well as factory personnel. Cost must be considered at all times along with function and reliability.

After the design is reasonably fixed, manufacturing must order tooling. With major new designs, the layout of the factory is often updated and rearranged. Purchasing must develop suppliers. Service must develop operator's manuals, parts catalogs, and training for dealer service personnel. Marketing must develop a plan for the introduction of the new product. Advertising has to be prepared. Pricing has to be determined and printed. Repair and service parts must be available when the first new machine is sold. These requirements have been present for a long time but their demands have grown with increases in size and complexity of machines.

So, although better technology has greatly improved the quality of new products being introduced today, it has not changed the total amount of time required from concept to birth of a new product. An obvious example is the Saturn car. Its birth came eight years after conception and six years after the first prototype was built.

Has Deere Tried Rotary Combines?

The roots of John Deere's rotary combine development can be traced back to some work done by Harvestaire, Inc., which was organized in San Francisco in 1954. They had developed a rather simple non-conventional combine. The threshing cylinder was replaced by a threshing fan with very high capacity. The straw walkers were replaced by a cyclone that was similar to those used on grinder-mixers but much larger and operated by a large suction fan. Consulting

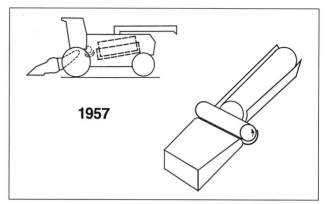

1957

Early experimental rotary combines used either an undershot or overshot feed from a conventional rasp bar cylinder to a rotary separator above grates similar to a concave.

engineers G.W. McCuen and W.L. Zink joined the team in early 1956. In turn they got an experienced combine engineer, Homer Witzel, on a one-year leave of absence from John Deere. With the Zink farm at Plano, Illinois, as the base of operations, over 100 tests were run in 1956 in California, Oklahoma and Illinois. The engineers became convinced that the threshing fan had potential but the cyclone was not a suitable way to separate and clean grain.

At the end of a year, Witzel returned to Deere with lots of ideas and great enthusiasm for eliminating straw walkers. By now it was obvious that a production machine was some time in the future so Witzel was assigned to the product development department of Deere & Company, headed by Charles S. Morrison. The department had a good track record for getting new ideas into production. They had done the early work for the John Deere Ottumwa Works on Deere's first twine-tie baler, the early work on corn heads for the Des Moines Works, and at the time were working with Ottumwa on the industry's first bale thrower.

The experimental combines from this department concentrated on changing only those parts of the combine that were known to limit capacity. Thus production parts were used for the header, feeder house, and cylinder. Straw walkers were replaced by a rotor for more positive feeding during separation. This permitted higher capacity at reasonable loss levels. The conventional shoe was also replaced, as it was subject to high losses in hilly terrain. The cleaning unit design was rather similar to the separating unit except that air was pulled up through the cleaning grates.

The product development department worked on a variety of rotary combines for more than a decade. Capacity was increased enough so that the John Deere Harvester Works started a parallel program, expecting it to lead to production combines. Complete success always seemed just a few years away. But rotary combines seemed to invariably require more power. They also made more chaff in dry conditions, resulting in a greater load on the cleaning shoe. Lastly, in tough straw or weeds,

the material would rope and plug the machine.

Deere's rotary combine experience from 1956 through 1980 was described in some detail by Kent Cornish, John Wilson and Don Yarbrough before industry engineers at the spring 1989 meeting of the Quad-City Section of the American Society of Agricultural Engineers. Their paper, "John Deere Combine Functional Development," showed nine designs which had been built and tested, in addition to the testing of all competitive rotary combines. The decision to stay with straw walkers in the 9000 series combines was based on many years of test experience with both types of combines.

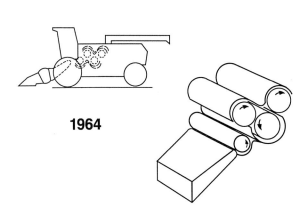

1964

Three separation rotors provided a serpentine path for the material after it passed through a rasp bar cylinder with closed-cell concave. This was a very compact design.

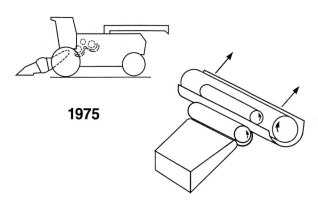

1975

A crosswise or transverse split-flow rotor was fed by a conventional rasp bar cylinder. This offered very good material flow but had complex drives and poor straw discharge.

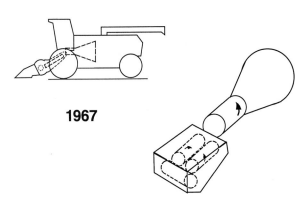

1967

The conical rotary separator had high capacity in corn and a huge grain tank. A rotary shoe wrapped around the conical separation unit.

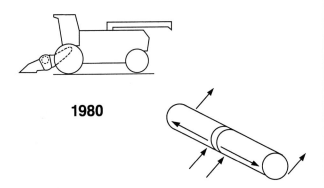

1980

Placing the transverse flow rotor in the header shortened the path for the straw. However, it lost the current flexibility of having a variety of headers for different crops with a common system for threshing and separation.

The 1169H SideHill combine offers the European small grain farmer with rolling terrain added capacity with reduced losses. The 1158, 1157, 1155, 1144 and 1133 complete the 1100 series.

Zweibrücken Cheers 25 Years of Combine Production

More than 90,000 combines have been built at Zweibrücken since production started in 1965. About 3,000 combines of 12 models are built each year by its 1,300 employees.

Grain is harvested in more than 70 countries by John Deere combines from Germany. Deere is one of the three top contenders for combine sales in the major European markets of France, Germany, the United Kingdom and Spain. The German-built 4435 combine is sold in the United States and Canada.

Other Overseas Milestones

John Deere, S.A. de C.V., Deere's affiliate company in Mexico, celebrated its 35th anniversary on November 10, 1990. In 1955, John Deere Mexico began building agricultural implements at its new factory in Monterrey. Eleven years later, in 1966, row-crop and utility tractors were added to its product line. A factory in Saltillo was acquired in 1974 and remodeled extensively for tractor production. The two factories currently make a variety of agricultural tractors, implements and components for Mexican and export markets, as well as buckets for John Deere industrial equipment. Earlier in July 1990, the new parts depot and the new product engineering center were opened in Monterrey. The product engineering center does design and development work for the Mexican market and some export markets.

John Deere is the number one agricultural tractor and implement manufacturer in the Mexican market. There are 95 sales and service outlets throughout Mexico. About 1,300 people are employed in Monterrey and Saltillo.

Deere & Company assumed full ownership of the company's affiliate in Australia in 1990. The name was changed to John Deere Limited, Australia and New Zealand from Chamberlain-John Deere Pty. Ltd. John Deere Limited employs about 400 people and is based in Perth, Western Australia. The factory in Perth builds tillage and seeding equipment designed for the Australian market.

The 200 John Deere dealers in Australia and New Zealand have made Deere the market leader in farm equipment sales in both countries. In addition to the equipment built locally, they sell tractors from Waterloo, Iowa, and Mannheim, Germany; combines from East Moline, Illinois, and Zweibrücken, Germany; and lawn and grounds care products from Horicon, Wisconsin.

Faster Cotton Picking

John Deere entered cotton harvesting in 1930 with the horse-drawn 30 cotton harvester. This and its successors were once-over boll pullers, now known as cotton strippers. Deere has been the sales leader in cotton strippers for many years.

Higher yielding cotton had traditionally been picked multiple times because the bolls ripen and open over a period of time. Thus a mechanical cotton picker had to be able to pick lint from the open bolls while leaving the green bolls and stalk undamaged for one or

Maximizer Combines Ready for the '90s

The 9940 cotton picker offered 90 percent more picking capacity than 2-row pickers. The large basket held enough cotton to allow picking a round of half-mile rows in 2-bale-per-acre cotton.

two future pickings. Deere entered the cotton picker market in 1950 with the 2-row No. 8 cotton picker from the Des Moines Works. This was followed in 1955 with the 1-row No. 1 cotton picker. This unit mounted on a 50, 60 or 70 tractor operating in reverse direction. Both of these pickers used a series of tapered and barbed rotating spindles that entered each side of the cotton plant to snag and remove the ripe cotton. Cotton was removed from the spindles by rotating, rubber-faced doffer plates. Air transported the picked cotton to the basket.

Over the next 30 years picking capacity increased with cotton yields but the acreage covered was limited to a 2-row machine working under 3 mph. In 1981 Deere introduced America's first 4-row picker, the 9940. This larger unit offered considerable savings in labor and fuel.

In the late '80s, service time to grease the picking drums each day was reduced by about an hour on 4-row pickers and about a half hour on 2-row pickers. The power dispens-

ing unit carried a week's supply of grease.

Deere introduced two additional safety features on cotton pickers in 1989. The "operator presence system" shuts off the picking units 5 seconds after the operator leaves the seat. A remote control allows the operator to stand on the ground and rotate the picking units slowly for inspection.

The new 9960 cotton picker was announced at the Palm Springs, California, show in January 1989 for sale in 1990. It introduced the most visible change in picking units since their 1950 origin. Conventional picker units pick from both sides of the row, using a drum on the right side of the row followed by a drum on the left side of the row. The new in-line units do all the picking from the right side of the row, with one drum following the other. The new in-line units were tested in Texas, Mississippi, California, Arizona

The 9960 cotton picker is the first one able to pick 30-inch rows. It is also the first 5-row picker.

and Australia. Picking efficiency matched conventional units in 1.5- to 3-bale-per-acre cotton.

The key advantage of the 9960 cotton picker is its flexibility for the farmer. Farmers have long recognized the yield advantages of narrower row spacings for corn and especially for beans. Until the in-line row units came along, it was not practical to plant cotton in 30-inch rows because there were no picking units that narrow. Switching row widths on the new lift frame is simple and fast, making it practical for a farmer to experiment with part of his crop planted in 30-inch rows and the remainder in traditional 38- or 40-inch rows. In addition to the previously described options, the farmer can set the in-line units for skip-row harvesting or mount 4-row conventional picking units for wide rows.

Another important feature of the high-capacity 9960 is the telescoping basket. This permits lowering the basket to 12' 6" for transport or entering buildings. In the field, the top of the basket can be raised 3 feet to provide 1173 cu. ft. capacity.

Deere has a significant export market for cotton pickers to several overseas nations. Most farm equipment is made in many countries but the high cost of tooling for cotton pickers has resulted in only two U.S. manufacturers and one in Russia. The Russian cotton picker uses a lower-cost, vertical spindle that is less efficient and thus not very competitive in the world market.

Deere's Executive Team for the '90s

A new Deere & Company executive team was named at the end of May 1990 with the retirement of Robert A. Hanson as chairman. Hans W. Becherer, who had been president and chief executive officer since September 1989, became chairman and CEO. Much of Becherer's 27-year career has been spent in assignments involving overseas operations, either in Europe or in Moline. He successively headed export marketing for the overseas

division; operations in Latin America, Australia and the Far East; and the overseas division itself. In 1986 he was elected executive vice president, worldwide agricultural equipment and consumer products.

David H. Stowe Jr. was elected president and chief operating officer. His most recent job had been executive vice president, worldwide agricultural equipment and consumer products division.

Hans W. Becherer brings to his job three years' experience as president, plus 24 years in other Deere assignments.

Bill C. Harpole was elected executive vice president, worldwide agricultural equipment and consumer products division. He had been senior vice president, worldwide parts and logistics division.

Thomas A. Gildehaus continued in his post as executive vice president, worldwide industrial equipment division. He previously had been executive vice president, corporate staff and administration.

This new team averages 53 years old. They are seasoned Deere managers with an average of 24 years of service. Their overseas experience provides them the background to keep John Deere a world leader in farm and industrial equipment. They have a total of 25 years working in Europe for Deere, and considerable additional time in the management of overseas operations.

David H. Stowe Jr. has considerable experience in worldwide industrial equipment and farm equipment.

1963–1990 Consumer Products

Starting with a single model, the 110 lawn and garden tractor in 1963, Deere became the world's largest manufacturer of lawn and garden tractors in less than 25 years. Deere now offers the most complete line of grounds care products available for residential and commercial users.

Lawn and Garden Tractors

Although there were earlier John Deere small tractors, it was not until 1963 that the Horicon Works in Wisconsin designed and produced a lawn and garden tractor for the homeowner, one that could mow lawns, blow snow, blade material, haul material and till the garden. The new tractor, designated the 110, was powered by a 7-hp Kohler engine and used a variable ground-speed control similar to that on combines. The power was increased to 8 hp for the 1964 model year.

With a 7-position lever to give increments of speed in each of the three forward gears and one reverse gear, and a single pedal to give infinite speed control plus clutching and braking, the tractor offered simple control as well as safer operation. Drive belts and rear wheels were shielded as added safety features. Deere introduced to the industry a safety start system on the ignition, ensuring that the tractor was in neutral before it could be started. This safety-start protection for inexperienced operators was adopted as a standard in 1972 by ANSI (American

One sales folder in 1935 featured the Model "B" tractor with special equipment for growing vegetable crops and flowers commercially. When the 9-hp Model "L" was developed in Moline, it too was suitable for garden or horticultural use.

The 110 lawn and garden tractor with 38" rotary mower established Deere in the rapidly growing consumer products field.

National Standards Institute). A fourth and lower forward speed was added in 1965 for tilling.

In addition, the company offered a front-mounted 36" snow blower, a 42" front blade, a front-mounted air compressor, and a No. 5 drawn and No. 7 integral sprayer. Approved equipment from outside manufacturers including an integral plow, planter, toolbar, disk, utility trailer, rotary broom, grille guard and weather shields expanded the adaptability and versatility of the 110 tractor.

In 1965, the 30 rotary tiller with optional right and left 8" extensions gave a full 38" width of cut in good soil conditions. The 80 dump cart, designed and built at Horicon, was also added to give the 110 tractor greater utility.

A hydraulic lift version, the 110H, was introduced in 1965. This feature provided power lifting of all attachments, making it even more fun to "play farmer" with the 110 tractor. During 1966

The fleet had grown to include one lawn tractor and four lawn and garden tractors. The 110 through 140 tractors were available in spruce blue, sunset orange, patio red or April yellow. But the optional colors were greeted with little customer enthusiasm and less dealer acceptance, so were dropped after two years.

other models, the 10-hp 112 and 112H tractors with Tecumseh engines, were announced for those requiring more power. Otherwise the specifications were the same as the 110 tractor.

Also for the 1966 season, Deere introduced a new product to the industry, the 6-hp 60 lawn tractor. The front-engine riding mowers that were on the market used light-duty transmissions and axles. The 60 lawn tractor had a heavier duty drivetrain, making it a durable unit for use with a front blade or snow thrower as well as a mower. Lawn tractors are generally simpler and lighter than lawn and garden tractors of similar power, and are not designed to operate tillage equipment for gardening.

First John Deere Hydrostatic Tractor Introduced

A second generation of lawn and garden tractors was introduced in 1967 with 540-rpm front PTO to improve versatility. The improved styling included a roomier one-piece fender deck, which was isolated to reduce vibrations.

It was late that year when the company announced the 14-hp 140 hydrostatic tractor, providing the consumer four choices from 6 to 14 hp. The H-3 version had an exclusive triple-func-

tion hydraulic system, giving independent fingertip control of front, center or rear-mounted equipment, most of which was larger for the new larger tractors. The H-1 version had a single control that could operate front, center or rear lift cylinders, while any equipment left mounted was held up by latches. An integral 3-point hitch was also available.

On both versions the clutch for driven equipment was electromagnetic. For the rotary tiller and similar equipment, the rear PTO ran directly off the engine, which was again a Kohler. Independent automotive-type rear brakes assisted steering, as on farm tractors. Dual headlights, taillights and a cigarette lighter were standard on the 140s, but were dealer-installed options on the 110 and 112 tractors. In 1968 an intermediate 12-hp hydrostatic model, the 120, was introduced.

Riding Mowers Added

The line was broadened again in 1969 when the first two riding mowers, the 5-hp 55 and 6-hp 56, were announced. The 55 used handlebar steering and had a 26" mower, while the 56 used a steering wheel and had a 28" mower. The 60 lawn tractor was upgraded to the new 7-hp 70. To

Consumer Products

For smaller lawns, riding mowers offered a lower cost way to ride while mowing than either of the two types of tractors.

round out the line, six walk-behind rotary mowers made by Toro were announced. A grass bagger was standard equipment on these 19" or 21" mowers.

With the above additions to the product line, along with those in the planning stage, the company committed the Horicon Works to consumer products exclusively. Grain drills were transferred to the Des Moines Works in 1969.

In 1970 the 7-hp 57 riding mower gave a third choice among these popular machines. A lower-cost 50 dump cart joined the original 80 dump cart.

Four chain saws purchased from Remington added a new market. They bore the John Deere name and included the normal after-sales support of operator's manuals, parts catalogs, and parts availability. The chain saws were provided by the OMP (outside manufactured products) division of Deere & Company in Moline.

The 90 electric riding mower was added in 1971, with one electric motor for traction and two for its twin-blade 34" mower. Three 12-volt batteries supplied the power to operate for about an hour, allowing time to cut nearly an acre. Its almost silent operation was a selling point, and like the gasoline-operated machines it could be fitted with a front-mounted blade for light work. Fifteen electric snow blower attachments were tested but were not sold, because the riding mower could not be slowed enough to work in

more than 3" to 4" of snow.

Deere was one of the first lawn and garden equipment makers to place full-page color advertisements in major national magazines such as *Life*, *Time*, and *U.S. News and World Report*. Ads promoted Weekend Freedom Machines, suggesting mowing Thursday or Friday after work so the weekend was free for recreation. This advertising program resulted in major increases in sales for Deere, and also helped competition by making people more aware of powered lawn-care products.

The 526, 726 and 732 walk-behind 2-stage snow blowers were a welcome addition for dealers in the snowbelt of the U.S. and Canada.

Recreational Products

The six main lines of consumer products introduced in the first eight years were all designed to make the work of the homeowner easier and more fun. This market was fitting for Deere, because all its consumer products were sold originally by farm equipment dealers. These products also made good use of Deere's engineering, manufacturing, marketing, product support and advertising skills.

84

The addition of rotary tillers in 1971 marked the entry into the third and last major walk-behind equipment market.

The slogan "Nothing runs like a Deere" aptly described the new snowmobiles made at the Horicon Works.

The scope of the consumer products line was broadened considerably with the introduction of the 400 and 500 snowmobiles in 1971 for 1972 sales. Deere now had recreational products designed purely for fun. Three more snowmobiles followed in 1972, including two JDX series finished in "Blitz Black," and another new green model, the wide-track 600. All five had 2-cylinder engines, from 25 to 40 hp. Two more snowmobiles, the 300 and JDX6, brought to seven the number of models available in 1973.

Early in the '70s, people in the U.S. and Canada were not only looking for the thrills of powered equipment like snowmobiles, but were beginning to invest time and money in fitness. John F. Kennedy had promoted fitness a decade earlier with little visible success. Deere decided to become a supplier of bicycles when the large market for adult bikes mushroomed in the early '70s. Bicycles had long been popular with adults in most other parts of the world for transportation and recreation, but in the U.S. and Canada, people graduated from bikes as soon as they got their car driver's licenses.

It was thought that bicycles would be good for all dealers, but especially helpful south of the snowbelt to provide a product to sell in the season of no mowing or gardening. Manufacturers in Taiwan were chosen as the source of the new bicycles, because of the well developed bicycle market there and because Deere did not have the facilities to manufacture their own. The Taiwanese bikes were redesigned and extensively tested at Horicon before production began. Unfortunately, the main Taiwanese manufacturers were not experienced in making lightweight bicycles, and their many subcontractors were not accustomed to consistently turning out high-quality parts. As a result, the bicycle venture lasted a mere three years, to 1976, because Deere could offer neither the quality of the name brands nor the price of the mass merchandisers.

Snowmobiles engineered and manufactured

His and hers bicycles were among the eight models introduced in 1973.

> *The concept of "Deere quality" is much easier to communicate in countries where John Deere products have been in wide usage for a number of years than it was in Taiwan. An example of incomplete communication was noticed on the first boxes of kids' 20" bikes from Taiwan. In large bold print was the familiar Deere advertising slogan, but a bit shortened to read, "Nothing runs Deere."*

at Horicon looked like a better business, and they did get added store traffic for John Deere dealers in an otherwise slow season of the year. Snowmobiles also helped expand the number of consumer products dealers. Two new snowmobiles, the low-profile 340 and 440, replaced the 300 and 400 in 1975.

The snowmobile line acquired exotic names in 1977—Spitfire, Cyclone, and Liquifire (a liquid-cooled machine). They achieved great success in the eight major cross-country races, coming in first in five and second in the other three. The result was that they were referred to as Big John and this became their trademark. It was only natural that as a follow-up a smaller, lightweight Little John model should join the line. Suitable color-coded snowmobile clothing was obtained from Hart Schafner & Marx for this very cool sport. This complete line of John Deere snowmobile clothing added considerable store traffic and sales volume.

Horicon clearly established that they could get into the recreational market and in a short period of six years have a product line well respected in the marketplace. Other products such as boats, off-road motorcycles and all-terrain vehicles were examined. The first two were already sold in well established specialty stores, and all-terrain vehicles were being sold with motorcycles. None appeared ideally suited for manufacture in Horicon, and the bicycle experience made some outside manufacturers seem questionable.

The Trailfire series of snowmobiles was introduced in 1978. Two more snowmobiles joined the line in 1979, the new Sportfire and updated Liquifire II. In 1981, the Sprintfire, Snowfire and Trailfire LX snowmobiles were announced.

Industry sales experienced a major decline as snowmobiles became a replacement market. Deere had invested enough in engineering and manufacturing to have a good product but had not learned how to make a profit on snowmobiles. The problem that plagued Deere and caused the shakeout in the industry was the unpredictable market for all snow products. If there are no heavy snows prior to Christmas, dealers and manufacturers are left with excess inventory that is often sold at a loss. Therefore, early in 1984 Deere reached an agreement with Polaris Industries of Minneapolis to take over the snowmobile line.

A New Tractor Line for 1974

Some 35 to 40 "allied equipment" items were listed in the 1973 catalog. Allied equipment was checked at Horicon and approved for compatibility with John Deere tractors before being authorized for sale. This equipment differed from that furnished by OMP in that both the equipment and repair parts were distributed to John Deere dealers directly by the manufacturer. Brinly-Hardy was the allied source for plows, disks and many other items.

Two skid-steer loaders, two electric chain saws and a line of hand tools, high-pressure washers, space heaters, battery chargers and hydraulic jacks carried the John Deere name. In addition chains, work gloves, fuel cans, flashlights, batteries and drop cords gave the consumer products dealer added diversity.

The big step forward in 1974 was the introduction of a completely new line of tractors with

> *All of the new lawn and garden tractors had lightbars in the front of the hood as part of the styling. There was a real dilemma in trying to decide if Horicon tractors should get this styling before the farm tractors did. It was finally decided to allow it since there were several other styling differences in the two tractor families.*

The 100 lawn tractor provided the homeowner the appearance of a farm tractor at an economical price for a mowing unit.

styling to match their big-brother farm tractors. The line started with the vertical-crankshaft Briggs & Stratton 8-hp engine 100 lawn tractor. The new lawn and garden tractor models were the 200 series with 8 to 14 hp in 2-hp steps, the 16-hp 300, and the top-of-the-line 400 with a 2-cylinder 19.9-hp engine. Power levels for consumer products are given in terms of engine horsepower while those for farm tractors are normally the more conservative horsepower at the PTO. All the lawn and garden tractors used Kohler 4-cycle gasoline engines.

On the new series all the engines were enclosed, ensuring quieter operation in addition to improved appearance. Earlier successful features like the forward-tilting hood for easy servicing of the engine and battery were retained. Mechanical lift was standard on the 200 series, with hydraulic or electric lift as options, and the controls were color coded for easy recognition and safer operation.

The 300 had all the 200 series features plus independent rear-wheel brakes and a fuel gauge on the gas tank filler cap. The 300 and 400 had hydrostatic drive. While dual hydraulics were standard on the 300, the 400 had triple hydraulics and front PTO as standard.

Power steering on the 400 was an industry first for lawn and garden tractors. It also featured high-flotation tires and a 2-speed rear axle, with the option of a rear PTO to provide power for a rotary tiller or similar equipment. A 3-point hitch was another useful option for use with plows, tillers, etc. A front loader completed the uses of this top-of-the-line tractor.

The previous chain saw line was replaced in 1974 by six new models made by Echo for John Deere. Two new Horicon-designed and built walk-behind mowers and an 18" trimming mower were added.

Three new riding mowers, the 66 and 68 (replacing the 56 and 57) and the 96 electric (replacing the 90), were announced in 1975. Two engine sizes were introduced in 1977 for the 300 series tractors, the new 12-hp 312 and the updated 16-hp 316.

Another riding mower, the 5-hp 65, was announced in late 1977, as was a new compact 216 walk-behind tiller which was destined to be in the line for a long time.

A Half-Million Lawn and Garden Tractors from Horicon

Another milestone was passed in 1977 with the production of the half-millionth lawn and garden tractor after only 14 years in the consumer products business.

With the growing line of Deere-manufactured equipment, the corresponding list of matched items from approved allied manufacturers was increasing to the extent that it was now divided into three categories—grounds maintenance, landscaping equipment, and gardening and weekend farming, with some additional units which could not be easily classified. A quick perusal of those listed revealed some 88 different items offered in 1978.

The No. 5 log splitter had been announced for the 1977 season and was joined in 1978 by five new chain saws, from the 30.1-cc Model 30 to the 70.7-cc 70V. A deluxe 21" self-propelled mower, the 21-SP, was added with a side discharge as standard. It had a 4-hp Briggs & Stratton engine and optional 2.6-bushel rear bagger. Another skid-steer loader, the 60, joined the 70 and 170 which had been introduced in 1973. Another product, a backpack power blower, was introduced.

Two new walk-behind snow blowers, the 8-hp 826 and 10-hp 1032, replaced the earlier 7-hp 726 and 8-hp 832, the last two digits indicating their width of cut in inches, while a compact snow thrower of new design, the 320, was announced for the coming season.

An attempt to offer lower priced lawn and garden tractors by eliminating fea-

tures in the new 208 and 312 models was not successful and was quickly abandoned. Introduction of the 312 required the 300 to be renumbered the 316, and a less deluxe 314 joined it. The 16-hp 216 tractor was added to the 200 line. In 1979, the 316 was replaced by the 2-cylinder 17-hp 317.

The first member of a completely new family of lawn tractors was introduced in 1978, the 8-hp 108. It was joined in 1979 by the 11-hp 111. The tremendous success of these models came from their bagging capability and the 11-hp Briggs & Stratton engine in the 111 tractor.

Yanmar Tractors Fill Power Gap

It has always been a challenge to meet the needs of all customers with the "right" tractor. The first 2-cylinder tractor to be called a John Deere, the Model "D," was obviously the right tractor for many customers as it remained in the line from 1924 through 1953. With 27 belt horsepower, it pulled three 14-inch bottoms where horses were limited to two 12-inch bottoms. The Model "D" was matched to the needs of large farms.

To cover more of the market, smaller tractors were introduced that were able to do additional farm operations beyond tillage and threshing. The ability to cultivate corn and other row crops

The experimental 101 tractor featured a rear engine, a mid-mounted seat for a good ride and excellent view to a front-mounted one-row cultivator, and an uncluttered platform.

The XR80 of 1974 was designed primarily for non-farm use. However, its rear engine, mid-mounted seat and open platform were reminiscent of the features found on the 101 tractor 30 years earlier.

accounted for the rapid acceptance of the Model "A," introduced in 1934 with 23 bhp (belt horsepower), and the Model "B" in 1935 with 14 bhp. The smallest John Deere farm tractor, with 9 bhp, was the Model "L," made in Moline by the Wagon Works from 1937 to 1946. Its engine was offset to improve visibility for cultivating single rows. Waterloo produced their smallest tractor from 1939 to 1947, the Model "H" with 12 bhp.

Theo Brown, manager of the general experimental department in Moline, felt he could design a better tractor for small farms. The all-important goal was to make a tractor ideally suited to cultivate one row. The first prototype was driven by Deere executives in January 1942. By January 1945, near the end of World War II, it was agreed that the 101 tractor was ready for final product engineering design, leading to production. The tractor was to have about 14 bhp, a 4-speed transmission, and be ideally suited for cultivating one row and plowing with a single 14- or 16-inch bottom. Implements were to be integrally attached to the tractor and hydraulically controlled.

The result of this further development was the Model "M," made in Dubuque from 1947 to 1952. It met most of the specifications laid down in 1945 with its 18 bhp, but had little resemblance to the 101 tractor. Rather, it kept the slightly offset engine of the Models "L" and "LA" it replaced and looked even more conventional than they did. The 2-row Model "MT" of 1949 took over the market vacated by the Model "H" in 1947. The rear-engine tractor was not dead experimentally, because two versions of a 2-row 202 tractor were in operation in 1950.

There always seem to be a number of customers who want a simpler, lower cost tractor. This was especially true in the southeastern U.S. where some horses and mules were still in use. In 1957, the product development department fitted or modified a plow, a cultivator and other implements to a small single-cylinder Lanz Bulldog tractor from Germany. When a phone survey was being made a John Deere dealer in the Carolinas retorted, "This is the fourteenth survey on small tractors. Just get us one we can sell!"

In 1960, Deere abandoned the market for farm tractors of 30 PTO hp and below, the main market of only a decade earlier. In 1970, Waterloo engineers recognized a void between Deere's largest lawn and garden tractors and the smallest farm tractor, the 820. They set out to design the ideal tractor for three main potential customers: the homeowner with 2-plus acres for grass mowing and snow removal, the commercial groundskeeper, and the part-time or weekend farmer.

The first XR80 was completed in June 1972 with an Onan 2-cylinder engine providing about 20 PTO hp. A hydrostatic drive was chosen, with each rear wheel driven by a John Deere radial-piston cam-lobe wheel motor. Cost targets were never achieved and the XR80 never reached production.

In 1976, Deere negotiated with Yanmar, the distributor of John Deere tractors in Japan, to supply some of their tractors to fill the price and power gap in Deere's North American market. The 22-hp 850 and the 27-hp 950, which were sold in 1978, were based on tractors that were well proved on Japanese farms.

Numerous modifications were made to meet the requirements of the market. Cooling capacity for the 3-cylinder diesel was increased 25% because parts of the U.S. experience much higher temperatures than in Japan. Ground speeds were increased over those needed in rice fields. An overrunning clutch was included in the 540-rpm PTO. Several changes were made in the seat and operator's station to match the larger size of customers in North America. A 2-post Roll-Gard structure was added for safety. Now the tractors were ready to receive John Deere Generation II styling.

The 850 tractor had the features, durability and 30 pieces of matched equipment needed by the part-time farmer. However, the majority of sales were for cutting grass.

Grounds Care Equipment

Under the overall umbrella of grounds care, a long line of utility tractors and loaders, backhoes, snow blowers and implements were grouped together with skid-steer loaders, lawn and garden and lawn tractors, riding and walk-behind mowers, and outdoor power equipment. The utility tractors in 1980 ranged from the 22-PTO-hp 850 to the 50-PTO-hp 2240.

The gear-drive 108 and 111 lawn tractors were joined in 1981 by a hydrostatic-drive 111 for the small acreage user, with one lever controlling forward-reverse direction and speed. At the same time a 16-hp 116 lawn tractor was also announced, with a 38" or 46" mower deck.

During 1982 the 18-hp 318 and 20-hp 420 with Onan 2-cylinder air-cooled engines joined the lawn and garden line, followed the next year by a new 16-hp 316 with similar engine. With front-mounted equipment or front weights to balance the heavier rear-mounted units, they increasingly performed tasks of their larger agricultural equivalents. Apart from the mid-mounted mower decks, the equipment too was imitating the larger machines, with rotary tillers, loaders, blades and snow throwers, to mention a few items.

The same year saw the Japanese-built "Taskmaster" line extended in each direction with the 14.5-PTO-hp 650, 18-PTO-hp 750, 33-PTO-hp 1050 and 40-PTO-hp 1250 tractors, the whole line having 2- or 4-wheel-drive option. Three options for cutting grass were offered with these tractors. The mid-mounted mowers provided the most uniform height of cut and the best maneuverability. Rotary cutters were the best answer for weedy areas that were mowed only a few times per season. The new rear-mounted grooming mowers provided the easy attachment of rotary cutters and nearly the mowing quality of the mid-mounted units.

For 1983 the riding mower and lawn tractor line was extended and updated. Five completely new riding mowers included two low-cost models, the R70 and R72 with 8-hp engines, and three deluxe machines, the S80, S82 and S92 with 8- or 11-hp engines, all five being Briggs & Stratton. Five nylon-line trimmers were introduced along

The first water-cooled diesel lawn and garden tractor made at the Horicon Works was only one of several newsworthy events in 1984.

with a new line of portable generators.

The new 3-cylinder 430 tractor was complete with power steering, single-lever hydrostatic drive with 2-speed rear axle, heel-operated differential lock, a turning radius of 26" and a new suspension system for the cushioned seat. The individual self-energizing brakes could be locked together for safer stopping. Options included a Category "0" 3-point hitch and a 2,000-rpm rear PTO. It brought diesel power and water-cooled engines to the lawn and garden tractor buyer.

On May 1, 1984, a 318 tractor became the one millionth lawn and garden tractor built at Horicon. Having taken 14 years to reach the half-million milestone, it had only taken seven additional years to achieve the million mark. Engineers from Horicon started kidding Waterloo engineers, saying that they worked for the John Deere factory that produced the most tractors. Deere sales of consumer products had doubled every five years since beginning.

A host of new machines were announced for 1984 including four commercial self-propelled mowers, two heavy-duty commercial tillers, two high-pressure washers and three space heaters. A one-bottom plow and the 44 Quik-Tatch loader were added for the 300 or 400 tractors. Weather

The F910 and F930 front mowers marked the beginning of a most successful line of grounds care equipment.

and 400 lawn and garden tractors. The gear-driven 850 and 950 compact utility tractors introduced in 1977 resulted in many loyal customers also. These tractors offered the advantage of better ground clearance, better loader operation, and optional mechanical front-wheel drive. Customers started asking for a combination of the best features found in both tractors.

Horicon Works engineers welcomed the clean-sheet design approach and made good use of their experience in the development of the 11-PTO-hp 655, 15-hp 755 and 19-hp 855, introduced in model year 1986. These tractors featured new styling that made them compatible with the gear-drive 650 through 1050 and certainly resembled their larger farm tractor cousins from Mannheim and Waterloo. They did not adopt the rear-engine design of the experimental XR80 of the '70s or the 101 of the '40s. Experience by Deere and the industry had shown that customers were willing to pay more for a lawn tractor than for a rear-engine riding mower, even though it had the same engine, transmission and mower. Customers prefer the appearance of a real tractor.

enclosures were now available for all lawn and garden tractors.

Responsibility for marketing the Yanmar-sourced compact utility tractors was transferred from Waterloo to Horicon in November 1984.

Included in the 1985 introductions were the 330 lawn and garden tractor and a very popular third front mower, the F935. A new 100 series of lawn tractors, the 130, 160, 165, 180 and 185, followed the pattern of the larger models with enclosed engines and new styling.

Horicon Designs Compact Utility Tractors

From the humble beginning of a single 7-hp 110 lawn and garden tractor, Horicon Works marketing, engineering and management had observed the trends in power levels and the use of hydrostatic drives. They had responded in 1974 with engine horsepower up to 19.9 and hydrostatic drive as standard on the new 300

The major design feature of the 855 tractor is its ease of use with various implements. Each of the three implements shown is easily mounted and removed, but for short jobs with one, the other two can be left on.

Rotary cutters are popular with customers who have land they wish to mow only a few times each year. Shown are the 503 economy, 506 regular-duty and 509 heavy-duty rotary cutters.

The new tractor line uses Yanmar 3-cylinder diesel engines rated at 16, 20 and 24 hp. Regular equipment includes 2-range hydrostatic drive, mid and rear live independent PTOs, planetary final drives, wet-disk brakes and power steering. Category 1 3-point hitch and mechanical front-

The AMT all materials transport, made at the Welland Works (Welland, Ontario) is an attractive alternate to a pickup truck in the field. It has a Kawasaki engine, variable-speed automatic transmission, and 4-wheel tandem drive.

wheel drive are optional. A wide variety of matching equipment is available.

In late August 1986, 12 golf course and sports turf machines were announced at a series of distributor meetings, giving the company another equipment field to develop. The Ottumwa Works was given responsibility for this line of products. Among the items shown were hydraulic gang mowers, the 90 boom mower, a 22" greens mower and five different aerators. Joining the line was a multifunction utility vehicle, the 1500 with 16-hp Kohler engine. This vehicle could be supplied with sprayer, spreader or utility bed attachments.

Enter the 3325 Professional Turf Mower

The number of machines offered in 1987 meant that the golf and turf purchasing guide consisted of 141 pages and the grounds care purchasing guide had 142, giving some idea of the complexity of coverage of these lengthening lines. One machine introduced that year which deserves particular mention is the 3325 professional turf mower. Built into the machine is a rollover protective structure, a full-size central grass catcher, single-lever reel lift, differential lock, cruise control, 20-gallon fuel tank, 2-pedal speed controls, and an instrument panel that monitors 12 machine functions. It has proved to be widely accepted since its introduction.

In 1987, the 150th anniversary year of the company also saw the riding mower line updated and two new front mowers added, the F912 and F932. The 14-hp 175 gave the lawn tractor line a sixth size. The new 332 lawn and garden tractor featured a liquid-cooled diesel engine.

The second generation of 200 series lawn and garden tractors joined the line in 1988. They featured durable overhead-valve Kawasaki engines. The two smaller tractors had 6-speed transmis-

The 3325 professional turf mower has a 38-hp Yanmar diesel engine. It provides an 11.5' cut with five 30" cutting units mounted on floating arms to better follow ground contours.

sion and the two larger had hydrostatic drive. A liquid-cooled gasoline engine powered the new 18-hp 322 lawn and garden tractor.

Mowing Equipment for the First-Time Buyer

By 1988 some 250 different machines comprised the consumer products lineup, but grass mowing remained the heart of the business. The challenge was to attract the first-time buyer as well as the experienced lawn and garden tractor buyer. Therefore, a new factory was built in Greeneville, Tennessee, to produce a new line of five 21-inch rear-discharge mowers with die-cast aluminum deck. About 5 million walk-behind mowers are sold by the industry in the U.S. each year.

For first-time buyers who want to ride, the company announced the all-new STX30 and STX38 suburban lawn tractors. These machines offered high value, with prices between riding mowers and the 100 series lawn tractors.

The Horicon Works also announced the F510 and F525 front mowers. They offer the homeowner the ultimate in visibility and maneuverability for mowing. The 14-hp unit has a 38-inch mower and the 17-hp unit has a 46-inch mower. A 38-

inch snow thrower is available for either model.

The 50 series gear-drive compact tractors were replaced by the 70 series from Yanmar with new clear-deck operator's station. The five models have fuel-efficient "TN" series 3- or 4-cylinder diesel engines with 18.5 to 38.5 hp. The 870, 970 and 1070 tractors offer planetary final drives, wet-disk brakes and continuous live PTO. Synchromesh 9-speed transmissions are optional on these three.

The Horicon Works also added a larger hydrostatic-drive compact tractor in 1988, the 955 with 33 engine hp. In July 1990, the company announced plans to assemble the 55 series compact tractors in a $20 million plant in Columbia County, Georgia. This was done to get the tractors nearer the major Southeastern market and closer to component suppliers.

The earlier AMT 600 all materials transport with handlebar steering was joined by the new AMT 622 with steering wheel, two seats and a 10-hp engine. A pedal-powered AMTeeny for the kids also joined the line.

Most mowing tractors are now equipped to collect grass clippings, from a twin-hopper 6.5-bushel unit through a 3-bag Power Pak to the more elaborate tilt dump and hydraulic dump systems with 13-bushel capacity.

The 285 and its sister lawn and garden tractors featured completely new styling and rapid hookup or detaching of equipment without tools.

A new departure in 1988 was the introduction of the 15 and 25 mini excavators, with 3-cylinder Yanmar engines of 14.5 and 23 hp. They complemented the larger industrial machines and gave the customer no less than 10 sizes to choose from.

The golf and turf line was extended with the 2243 professional greens mower, updated vacuum sweepers and a new 10-hp 1200 utility bunker and field rake. Two allied products appeared in this line's catalog for the first time, the Broyhill series of pull-behind sprayers and the Coremaster 12 aerator.

It was August 1990 before another extensive announcement of over 25 new products was made in Orlando, Florida. Among the new machines was another transport, the AMT 626, four LX lawn tractors, four GX and SRX riding mowers, and two lower-priced walk-behind mowers.

Commercial users got two new mid-size front mowers, the F710 and F725. Also new for

Kohler 9- and 12.5-hp engines power the STX30 and STX38 tractors, which have 30- and 38-inch mowers respectively. The 5-speed in-line gearshift is fender mounted.

grounds care were the Deere-designed, Deere-manufactured 48- and 54-inch walk-behind mowers. They featured antiscalp oscillating decks with single-handle height control.

Two push-type and three self-propelled mowers are available, powered by 2-cycle 4-hp or 4-cycle 4.5-hp engines.

1960–1990 Industrial Equipment

Starting with adapted agricultural tractors in the '20s through '50s and moving to a broad line of crawler dozers and loaders, backhoe loaders, scrapers, motor graders, log skidders, 4-wheel-drive loaders, excavators and other specialized machines, the industrial equipment division is now a leader in world markets.

In the Beginning...

It was inevitable that a manufacturer of agricultural equipment would find occasions when that equipment had industrial applications. This has already been illustrated in the first volume of this work with the Waterloo Boy tractor pulling a city road grader in 1921 in Moline, Illinois, (p.29), and various versions of the Model "D" adapted for industrial use (p.31 and p.109). It could be argued that at an even earlier date the log wagons sold by the Portland branch in 1909 indicated this beginning.

Deere Archives show other applications of the Model "D" on both solid and pneumatic rubber tires. It was 1935, however, before the industrial Models "DI," "AI," "BI" and later the "LI" were officially announced.

The commitment to these industrial versions of the agricultural base machines was not large—about 100 of the Model "DI" were built from 1935 to 1941, for instance. We must come forward to the postwar years for signs of real progress in this field.

A similar but more extensive use of the basic farm unit was the supplying of Model "BO"s without axles to Lindeman Power Equipment Company of Yakima, Washington, as the basis of their small crawler tractor.

In December 1946 the company acquired certain assets of the Lindeman company and initially continued to build the same BO-Lindeman crawlers. In June of that year the foundations were laid for the new Dubuque factory, which was destined to become the company's chief industrial facility.

An interesting application of the Model "D" appeared in 1928 as the motive power for the Hawkeye Motor Patrol, one of the earliest attempts at a motor grader.

The popularity of the Model "BO" tractor in Washington orchards led Lindeman to provide a crawler version for stability and traction in hillside orchards.

John Deere's First Crawler

The Dubuque Works was originally constructed to build the "M" series of tractors. It was inevitable that the Model "BO" was soon replaced with the "M," and the "MC" crawler was the result. To begin with, the base units were shipped from the Dubuque Works to Yakima, Washington, to have crawler drive and tracks added, but this was uneconomic and the whole operation was transferred to Dubuque.

Their use in woods resulted in an advertising slogan, "The woods are full of them," and Deere's position as the leading supplier of orchard and forestry machines could well have started from this small beginning. By May 1952 the first three items of allied equipment had been approved for use with the Model "MC"—two winches and a posthole digger.

In the summer of that year the "M" series was updated to become the 40 series with an additional 2 hp. More important than the increase in

power was the adoption of the 3-point hitch. This Category 1 size hitch was standardized by ASAE seven years later. The hitch greatly increased the variety of machines matched to the tractor, with several more machines officially approved. The 40C continued to use the 3-roller track frames of the Model "MC."

During the period from 1952-54 the first Deere-produced attachments were announced, with the 61 angle-dozer becoming very popular with loggers. It was followed by the 80 rear blade, 20 scoop, TP26 forklift, 47 mower and LFI road maintainer.

The introduction in late 1955 of the 420 series, with the crawler having the choice of either a 4- or 5-roller track and a further increase in engine power to 30 hp, produced a machine with real industrial potential. A 5-speed transmission and direction reverser further emphasized this.

As a result 1956 saw the rapid expansion of approved allied equipment to some 50 items, from loaders to backhoes to scrapers. Suppliers such as Henry, Davis, Shawnee, Gearmatic and Hancock built machines power-matched to the new line of tractors. These were all shown at the 1957 Chicago Road Show, the first and most colorful of all road shows. The majority of tractors were still painted green and yellow up to this time.

Most of the industrial equipment activity was based on Dubuque tractors, but contractors had started using Waterloo's rugged and powerful 80 farm tractor for pulling Hancock elevating scrapers.

The industrial division gained its identity separate from the farm machinery business in 1956 when separate records of industrial sales were kept for the first time. The first dealer to exclusively sell industrial equipment was appointed in Ft. Dodge, Iowa. A marketing organization for the United States was established in 1957 with two

The 5-roller 420 crawler with Ateco loader and ripper marked the real beginning of the industrial line.

assembled. The Wagon Works was converted to industrial products in 1958 and carried the new title John Deere Industrial Equipment Works.

The 440 Wheel and Crawler Tractors

In 1958 the first tractors designed for industrial use came off the assembly lines at Dubuque. Painted "industrial" yellow, they were the 2-cylinder 440 wheel and crawler models with the option of a John Deere gasoline engine or a General Motors diesel.

Also introduced were the 831 loader for the crawler, 70 and 71 loaders for the wheel unit, and 50 (center mount) and 51 (5-position) backhoes for both. The 300 sideboom and several other attachments were also added to the lengthening line. The 831 crawler-loader, as the unit was referred to, became very popular in the U.S. and Europe, with or without a backhoe. The tractors assembled in Mannheim had Perkins 3-cylinder diesel engines instead of the GM.

The year would also be remembered for the introduction of the first all-hydraulic bulldozer, the 64, which was an instant success. It was the first of many innovations introduced subsequently.

regional sales branches. Also in 1957 an engineering department was started with John L. French from the Harvester Works as its head, plus three engineers from the Spreader Works who had been working on attachments for industrial tractors. An office was established in the basement of the Wagon Works, where a grandson of John Deere once had the Velie automobile

Waterloo offered two versions of the 820 tractor for use with Hancock scrapers. The model with the offset operator's station provided improved traction and maneuverability.

The backhoe loader has become Deere's best selling industrial product. The 51 backhoe shown here, introduced as an attachment in 1958, could be offset right or left to dig next to walls.

John Deere industrial equipment pioneered multiple-function control levers. The two controls on backhoes provided four functions, and the T-bar control on dozers controlled height, angle and tilt. This approach was taken a step further in the Pilot-Touch single-stick crawler control. This option provided hydraulic operation of the direction reverser, master clutch and clutch-brake steering. On the main competitor's crawlers, these functions required four hand controls plus two foot controls. Unfortunately, the poor reliability of this feature in the field led service to convert these crawlers back to conventional Deere controls.

Late in 1958 the farm tractor styling changed from the 20 to the 30 series. At the same time the largest model, the 75-hp 830, was joined by a special industrial version, the 840. Its offset driver's position, previously tried on the 820, allowed the new 400 scraper to be attached by a gooseneck. The 230 cable plow and 800 log fork were also added to the line.

Dallas Announcement Affects Industrial Line

Deere Day in 1960 initially brought 4-cylinder engines to the smaller industrial models, the 440s being replaced with the 40-engine-hp 1010 series, both wheel and crawler. This was primarily an update in engines, with limited changes in the industrial-version tractors. Later in the year the all-new 2010 wheel and crawler models appeared. Also added in 1960 were Deere's first winches, the 10 and 20 for use on the 1010 and 2010 crawlers. Deere also introduced the first hydraulic direction reverser for crawlers.

Gasoline or diesel engines were available in all models, including the new and much larger 3010

"Look...the Steering Levers are Gone!"

Operators loved the single-stick control option on 440 crawlers—when it worked.

99

The much heavier loader boom matched the increased power and weight of the new 52-hp 2010 crawler.

and 4010 industrial wheel tractors. These new tractors were the first in the industry to use planetary final drives and wet-disk brakes.

During 1962 the John Deere Industrial Equip-

Customers had to wait several years before they could have the muscle available in the experimental 8-speed C-1 crawler (3010).

ment Works, which built backhoes, loaders, sidebooms and blades, was doubled in floor space. Construction also began on a Product Engineering Center in Dubuque.

Enter the "JD" Line

The latter part of 1962 saw the replacement of the 830 with the new 127-hp 5010, and the following year the 840 scraper became the 5010 scraper. New 93 and 95 backhoes replaced the popular 50 and 51 in 1964, while a truck-mounted pulpwood loader, the 345 Rotoboom, was introduced. Also in 1964 the 3010 and 4010 industrial tractors were replaced by the JD500 and JD600, and the 5010 equivalent became the JD700.

The new identification system developed had 3-digit numbers, the first digit indicating size, the second classification, and the third model configuration, all preceded by JD and followed

The JD450 and JD350 crawlers set new standards of productivity for small crawlers and gained a high share of the market.

designs continue to provide John Deere industrial machines that look clean, rugged, and functional.

The third element of importance for this time was the birth at the Dubuque Works of the 300 series engine, which continues to be Deere's most popular engine world-wide. It had been developed for use in the worldwide 1020 and 2020 farm tractors. However, it was well adapted to a broad spectrum of industrial equipment. This engine in its 3-, 4- and 6-cylinder versions provided much greater reliability and offered considerable room to grow in power.

In this mode, the new JD350 and JD450 crawlers were classed as either dozer or loader models. Wheel tractor models were replaced by the JD300, JD400 and JD700-A. Equipped with a 9-cu.-yd. scraper, the JD760 replaced the 5010, which had set a high standard in its class.

Allied equipment was still used extensively in 1965 to broaden the applications available. This same year it was Dubuque's turn to have a large sum spent on its expansion, and at the same time the

by -A, -B, etc. as the models were updated. This system was continued until 1980, when the JD and hyphen were dropped.

An investment of $40 million resulted in a new line of basic tractors for 1965 and a significant change in concept. Perhaps more significant than the new model numbers was the change of thinking. Up to this time industrial equipment had been looked on as tractors with attachments; now each machine was considered an integral unit. Henry Drey-fuss Associates had a major input into this inte-grated appearance. Their

The JD300 tractor with loader and 92 backhoe was a popular combi-nation. Its designers built on previous experience with backhoe loaders and combined that with a much better tractor.

The new JD480 forklift was based on the JD400 tractor. This rough-terrain forklift was especially productive on construction sites.

World's First Articulated-Frame Motor Grader

Another major step forward in 1967 was the introduction of the world's first articulated-frame motor grader, the 83-hp diesel JD570. Frame steering offered two main advantages. By bending in the middle, the motor grader had a much shorter turning radius for working around street corners. On the straightaway this design let the operator offset the front wheels, to avoid windrows or keep the rear wheels on better footing. A differential lock could be engaged to reduce spinning in normal operation, or released for shorter turns.

A larger skidder, the JD540, was added in 1968 with diesel engine only.

A seventh product line was added in 1969, the 131-hp JD690 excavator with Hitachi undercarriage. This entry gave Deere the potential for eventually covering 80% of the industrial mar-

production of bulldozers and loaders was transferred from Moline to Dubuque. One other new line was also added, a JD600 "Pipeliner Special" for pipelaying contractors.

The JD440 skidder marked Deere's serious entry into forestry equipment in 1965.

The JD570 motor grader marked the biggest improvement in design since the first self-propelled grader of 1926. It featured an articulated frame, all-hydraulic blade positioning, and the grader industry's first Power Shift transmission.

ket by type and size. The JD644 loader plus a 152-hp JD760-A scraper and another even larger model, the 15-cu.-yd. JD860 with 215-hp engine and only four wheels, opened up anoth-

By introducing the JD544 4-wheel-drive loader in 1968, the sixth of the product fields envisaged in the early '60s had been entered. The JD544 was articulated and originally offered with a choice of 94-hp diesel or gasoline engine.

er dimension in the construction sphere.

The updated JD500-B backhoe completed the list of machines shown at the CONEXPO '69 equipment show, and was evidence of the company's growing strength in the industrial equipment world.

Enter the '70s

The rapid growth of industrial equipment sales led Deere & Company to reorganize its management structure into three divisions in 1970: farm equipment and consumer products, United States and Canada; industrial equipment; and overseas. In 1971 the industrial dealers' agreements were divided into construction, utility, and/or forestry, which was a natural progression as more machines in the line allowed increasing specialization. Industrial equipment sales hit $217 million in 1970, $306 million in 1972 and $465 million in 1974.

The '70s saw the crawlers become the B series,

103

Industrial Equipment

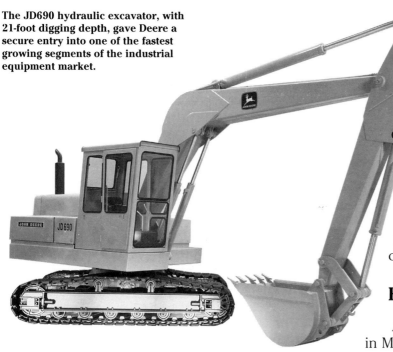

The JD690 hydraulic excavator, with 21-foot digging depth, gave Deere a secure entry into one of the fastest growing segments of the industrial equipment market.

the JD350-B with a 42-hp engine and the JD450-B with 65 hp. New introductions included a 27-hp skid-steer loader, the JD24, and two small trenchers, the JD11 and JD21. The Welland Works in Ontario provided two knucklebooms for loading pulpwood, the 3805 and 3807.

In 1971 two smaller utility tractors, the 43-hp JD301 and 59-hp JD401, joined the yellow line along with two forklifts—a smaller JD380 and an updated JD480-A. There were also four new integral backhoe loaders announced, the 50-hp JD310, 62-hp JD410, and 80-hp JD500-C and JD510.

Not until 1973 was the larger JD740 skidder added to the forestry line. It featured a 2-winch option, with the lighter winch retrieving individual logs and the main winch towing the accumulated pile of logs. A galaxy of other new machines joined the line that year, including the 9500 sideshift backhoe; JD301-A, JD302/JD302-A, JD300-B, and JD401-B/JD401-C utility wheel tractors and loaders; and the JD450-C crawler.

The new, larger JD555 crawler loader was announced in 1974, with Deere's first torque converter in a crawler. The JD646-B compactor

was derived from the JD644-B loader to work in sanitary landfills.

Products from the 13-year-old original Industrial Equipment Works in Moline were transferred to Dubuque, and the former industrial facilities were taken over by the consolidated Plow & Planter Works. The Dubuque Works was expanded by more than a million square feet in 1974 and a further 800,000-sq.-ft. extension was started. In 1973 Deere acquired 960 acres of land near Davenport, Iowa, for a new factory for the industrial division. Production commenced in December 1974.

ERA III Announced

At a worldwide industrial sales meeting in May 1974, with 28 different models of the various categories as its base, the company announced the ERA III program. ERA I had been the period leading up to the '60s, and ERA II the consolidation of Deere as one of the top five or six suppliers of industrial equipment.

ERA III was the announcement of the machines Deere intended to introduce over the 5-year period of 1975-1979—what a milestone to talk so frankly with dealers about the coming five years. These plans included a large excavator, four more graders, a large 4-wheel-drive loader, an improved scraper, and four crawlers with hydrostatic-drive transmissions. Experimental models of many of these proposed products were on display. This announcement was without precedent in either the company's or the industry's history.

As the company's president, Ellwood F. Curtis, stated at the time, "We are at an important jumping-off point—probably the most important in the short history of our Industrial Equipment Division. We're convinced the time is right for a big move forward."

The machines that contributed to this expansion during this "golden age" were the JD550 bulldozer; JD750 dozer and JD755 crawler load-

The 750 crawler dozer heralded Deere's entry into the large crawler class. The Dual-Path hydrostatic drive with independent track control provided power on both tracks during turns for excellent maneuverability and maximum productivity.

The JD743 tree harvester clamps and shears the tree, then lays it between spiked rollers. Cylindrical knives close around the trunk and remove branches as the rollers propel the tree to the rear.

er; JD670 and JD770 motor graders; JD640, JD540-B and JD440-C log skidders; JD310-A backhoe and JD444 4-wheel-drive loader, all introduced in 1975-76.

The JD693-B feller-buncher provides good mobility in the forest. The tree is clamped and then sheared at its base.

The following year the JD693-B feller-buncher appeared, based on the excavators of the JD690 series, together with the JD743 tree harvester which had been developed in the early '70s. As scheduled for 1978, the big JD890 excavator, JD860-B scraper, JD670-A and JD770-A graders, and the largest crawler in the shape of the JD850 dozer and JD855 crawler loader filled out the line.

Finally in 1979 the program envisaged five years previously was completed with the introduction of the JD750 widetrack bulldozer and the JD844 articulated loader. The previous year the gasoline engine option available on some utility products was discontinued.

The Hydraulic and Electronic Era

Hydraulics played a major part in these new machines. The two graders announced in 1978 had hydrostatic front-wheel drive systems, and electronic sensors that could detect slippage of the rear wheels and engage the front-wheel drive automatically. Other examples where hydraulics were applied were the Dual-Path crawler drives, scraper elevator drives, and motors to propel excavators and skid-steer loaders.

105

The hydrostatic front-wheel drive of the 672A grader provided better steering and traction.

The introduction of a new scraper, the JD862, in April 1980 emphasized that the machines envisaged six years previously were not to be the end of the expansion of the line. Rather, the JD862 signaled the beginning of the company's high-tech contributions to the industry. Its controls were no longer levers but rocker switches in the padded armrests of the seat. More advanced than this was a black box with microprocessor controlling six sensors to determine the best shift sequence, shift the transmission automatically, control the torque converter, protect the whole mechanism and diagnose faults. The aim was to ensure that productivity would be 95% of maximum. Another newcomer in 1980 was the JD740-A skidder.

Productivity of the 862 scraper was increased by operating at optimum capacity all the time. Electronics assisted the decision-making of the operator, just as hydraulics had earlier reinforced his muscles.

Houston hosted the CONEXPO show early in 1981, and Deere took the opportunity to introduce a number of new machines and at the same time adopt a new style of identification. The "JD" was dropped and the name John Deere in larger size was now in attractive decals in place of the earlier chrome and black plate.

The new models included a scraper, the 762A; two hydrostatic crawler loaders, the 655 and 755A; and the 4-wheel-drive 444C, 544C and 644C loaders and 646C compactor. The full industrial line of 68 models included 29 utility, 22 construction and 17 forestry machines as well as various attachments and allied equipment.

A Worldwide Industrial Equipment Organization

Following the ERA III announcement in 1974, Deere formed a worldwide industrial equipment marketing organization, John Deere Intercontinental Limited S.A., with an office in Brussels to cover Region II serving Europe, the Middle East and Africa. This was followed in 1978 with the splitting out of industrial equipment from JDIL in Moline to better serve the Region I markets of Latin America and the Far East.

The John Deere Training Center underscores Deere's commitment to training by providing hands-on experience to over 3,000 salespersons, service technicians and customers each year.

By 1977 sales went past the half-billion dollar mark, rising by 48% that year to $670.1 million, and in 1979 reached a new peak of almost one billion dollars, $996.8 million.

By 1981 some 674 dealer outlets worldwide serviced this growing line, 234 of these overseas, and with the addition of the larger 800 series machines the company now had coverage of over 80% of the world market.

By this time both the Dubuque and the Davenport factories had doubled in size in the previous seven years and now had 168 acres under roof. In addition, plants in Mannheim, Germany, in Saran, France, and in Perth, Australia, were all producing industrial equipment.

One other important and significant event took place in February 1981, the opening of the John Deere Training Center in its new 55,000-sq.-ft. headquarters near the Davenport Works. This center provided the backup near the corporate headquarters necessary to support both the industrial and agricultural equipment divisions anywhere and anytime. Field demonstration and training facilities for industrial equipment at Coal Valley, Illinois, and for forestry equipment at Rome, Georgia, joined the Phoenix, Arizona, field demonstration and training site that had opened in 1978.

Robert J. Gerstenberger had a major influence on Deere industrial equipment, having spent 27 of his 31 years with the company in that area. From industrial sales manager for the Portland sales branch in 1955, Gerstenberger progressed to director of industrial equipment marketing at Deere & Company in 1966 and to senior vice president, worldwide industrial equipment division in 1977. Upon Gerstenberger's retirement in July 1982, his post was taken by David H. Stowe Jr., who had been a familiar figure in Europe and was destined to become the company's president in 1990.

A Galaxy of New and Modified Models

Products announced for the 1982 season included three excavators, the new 990, the updated 890A and a wheeled 690B for greater mobility; the 310B backhoe loader; the 750 narrow-gauge dozer for work in oilfields; and the 855 steel mill loader, which could handle molten slag.

Over the next four years the line continued to be updated. In 1983 four backhoe loaders were added, making a total of seven. An agreement with Hitachi in Japan extended the excavator models offered to six, while the forestry line was augmented with five new log skidders and another feller-buncher.

All the crawler and 4-wheel-drive loader models, as well as the two smaller log skidders, were

The top-of-the-line 990 excavator featured a John Deere 260-hp V-8 engine along with increased digging depth and capacity.

updated for the 1985 season, but 1986 was to see the biggest modernization program. No less than 22 machines were involved and in addition three new mini excavators and a new wheeled model, the 595, were added.

The biggest news in the 1986 introduction was the all-new C-series backhoe loaders, from the 55-net-hp 210C to the 115-hp 710C. Designers were given a clean sheet. These tractors had little similarity to the farm tractor line except for the engines. The unitized mainframe provided isolated mounts for both the engine and cab to minimize noise and vibration. The controls, comfort, convenience and view provided by the new operator's station are said to be the best in the industry. Engine, transmission and rear axle can be removed independently for easier overhauls.

All C-series backhoe loaders feature a torque converter, power-shift reverser and differential lock. Mechanical front-wheel drive is optional.

With more dealer outlets and a modern line of equipment, the industrial division looked forward to further expansion in the late '80s. Progress toward the objective of "total waste elimination," to reduce costs throughout the manufacturing process, was paying dividends.

The company's 150th anniversary in 1987 nearly coincided with the first 40 years of industrial equipment, and an additional 21 new or improved products were announced. While Hitachi had been supplying some of the excavator models since 1983, Deere was now sending backhoes and 4-wheel-drive loaders to Japan for Hitachi to sell under its own brand name.

The increased reach of the E-series 4-wheel-drive loaders makes it easy to load trucks and trailers with gravel or dirt.

The Billion Dollar Sales Mark Passed

In 1988 the company sold $1.162 billion of industrial equipment worldwide, thus passing the previous peak reached in 1979 before the depression of the '80s. Operating profits for industrial equipment passed the $100 million mark for the first time in 1989. Industrial equipment continued providing over 20% of net equipment sales for Deere, an achievement first reached in 1972.

The expected growth in the small- and mid-size machine market had occurred and Deere's strategy in concentrating on these segments was amply justified. The backhoe loaders introduced in 1986 are quite competitive in a North American industry market that grew from 11,000 units in 1982 to 24,000 units in 1987. The new G-series small crawler dozers and loaders in 1988

focused on the arena that led Deere into the industrial equipment business. Deere has been a leading contender in crawler dozers, a North American industry market that has grown to over 12,500 units from only 7,000 in 1982.

Also new in 1988 was the E-series 4-wheel-drive loaders and log loaders. This all-new design featured Z-bar linkage to the bucket, giving greatly improved control and performance, together with wider boom arms that gave better visibility, as did the larger, roomier cab, and a new bucket designed for faster filling and better load retention. Added to these improvements were a wider mainframe and longer wheelbase, making servicing easier and improving the ride characteristics. This was followed by the new top-of-the-line 216-hp 744E 4-wheel-drive loader in 1990. Its exclusive downshift system provides more push while filling the 5-cubic-yard bucket, and more speed during maneuvering.

The fastest growing segment of the industrial equipment market has been hydraulic excavators. North American industry sales quadrupled in six years and stood at just under 10,000 units in 1988. Deere had started supplementing their excavator line in 1983 by importing Hitachi-built units with John Deere engines. Another joint venture with Hitachi started when a new factory in Kernersville, North Carolina, began producing excavators in 1989.

A 495D wheel excavator provides rapid mobility on road jobs. It is setting concrete barriers near John Deere's original roots in Moline, Illinois.

PART II

PRODUCT REVIEW

Tractors, United States and Canada

The 10 Series, a New Generation

In the early 1950s, farmers were asked about their future requirements. They requested more power to farm more land in less time. Other requirements were more power at the PTO, bigger fuel tanks to give longer working days without refueling, better transmissions, hydraulics that would do more, and more operator comfort with easy to operate controls. All these features had to be available at a price the farmer could afford. Effectively this meant that the design team had to start from scratch.

For nearly seven years prior to 1960, John Deere had experimented with and fully tested the New Generation tractors. There were four new tractor models, the three smaller ones with 4-cylinder engines, the larger with six cylinders. The new models were the 1010, 2010, 3010 and 4010. The latter, in its diesel version, would become the most copied tractor in the industry's history. The advertised PTO horsepower of these new models was 35, 45, 55 and 80.

Hints of the new tractor styling had been seen in the last series of 2-cylinder tractors—the sloping automobile-style steering wheel and instrument panel, and rear fenders (for the row-crop models) with built-in lights. Nevertheless, 95% of the parts used in the New Generation tractors were newly designed.

The list of new features for the two larger models included comfortable, scientifically designed suspension seats, exclusive hydraulic power brakes, and 540/1000-rpm rear PTO and 1000-rpm mid PTO for new machines being developed. At a stroke Deere had jumped years ahead of the competition. The 1010 and 2010 were built at the Dubuque Works, the 3010 and 4010 at the Waterloo Tractor Works.

The 1010 single-row-crop, shown here with No. 1 crane, is the obvious descendant of the 430 standard.

The 1010 special row-crop utility was a low-cost basic tractor. The pan seat is like those on letter-series tractors of the,'40s.

The 1010 series grew to include the row-crop utility and tricycle (front row); special row-crop utility diesel, row-crop with adjustable front axle, and utility (middle row); and the single-row-crop and 4-roller crawler (back row).

The 1010 utility is shown with rear exhaust and the standard seat cushion, a carryover from its 2-cylinder predecessors.

The 1010—the 330/430 Replacement

The 1010 was initially offered in single-row-crop (formerly known as standard), utility and row-crop utility versions, the same as its predecessors were. Agricultural crawlers painted green and yellow were available with 3-point linkage and 4-or 5-roller track frames.

A low-priced version of the 1010, the special row-crop utility, was available with pan seat, 540-rpm PTO, and 5-speed transmission, to compete with other small tractors on the market. The rest of the 1010 wheel models had independent "live" 540/1000- rpm PTO. All the wheeled 1010s had 5-speed transmissions; the crawler had a 4-speed. All could have gasoline or diesel engines.

The 1010 tricycle, shown with rear exhaust, was a worthy successor to the 430 tricycle and the original Model "MT."

This 1010 agricultural crawler has the base 4-roller track frame, 3-point hitch and reduced-height muffler. It would pull a 4-bottom plow in most conditions.

The 2010—a Small Tractor with Big-Tractor Features

The 2010 series tractors were styled to more closely match the Waterloo models. Fitted with the same type of fenders, scientifically designed deluxe suspension seat and dashboard controls, they were smaller versions of the two larger tractors in most respects other than the engines. Available for gasoline, LP-gas or diesel, the

This 2010 row-crop tricycle tractor reflects the popularity of LP-gas in the gas-producing areas.

The 1010 grove and orchard tractor was built low—just over 4 feet to the top of its hood, and suitably streamlined to prevent damage to trees. This model was derived from the 1010 utility.

A 2010 row-crop with adjustable front axle carries a 1010 row-crop utility on a grain drill to illustrate the strength of the drill frame.

4-cylinder engines had a sleeve-and-deck design similar to the 1010 series, providing compactness and efficient heat dissipation. In these "over-square" engines, the bore was greater than the stroke.

In the fall of 1962 the 2010 series were improved with a lower range of transmission speeds, easier rear wheel adjustment, easier servicing, faster power steering, and a new front rockshaft with down-pressure. Besides the choice of engines, the row-crop model offered Roll-O-Matic, dual-wheel, single-wheel or adjustable-axle front ends. For farmers who needed to change rear wheel tread frequently, power-adjusted rear wheels were an option. Those preferring a lower tractor could opt for the row-crop utility, which offered the choice of straight or swept-back front axles.

Other models in this series were the 2010 Hi-Crop and (in 1962) the 5-roller 2010 agricultural crawler. All 2010s, including the crawlers, had an 8-speed Syncro-Range transmission, with on-the-move gear changing in each of four ranges. They could all be equipped with the new Quik-Coupler, first seen on the 8010, for hooking up implements without leaving the tractor seat.

A 2010 row-crop utility is disking with a mounted "U" rigid tandem disk.

The single front wheel of the 2010 reduced damage to a cotton crop harvested with the 277 mounted cotton stripper.

The low-cost, diesel-powered 2010 special row-crop utility with 25-B unit planters. Each dry-fertilizer hopper held 160 lb.

A 2010 Hi-Crop tractor with front-mounted cultivator and front rockshaft uses its high clearance as it cultivates a tall crop.

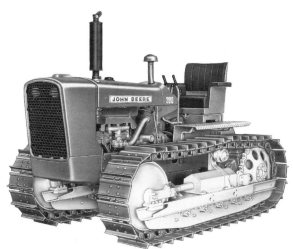

The 2010 agricultural crawler provided more power and better features than previous John Deere crawlers.

The operator of a 2010 diesel row-crop utility tractor stops to check with the operator of a 3010 diesel row-crop utility tractor with a disk.

The 3010—the Largest 4-Cylinder Model

The 4-cylinder version of the famous 6-cylinder 4010, the 3010, could easily handle a 4-bottom plow or 4- and 6-row planting and cultivating equipment. Offered as a row-crop, standard, row-crop utility, or grove and orchard tractor, it proved ideal for the medium size farm.

The 3010 tractor—along with the 4010—set design standards for many years with features like the Syncro-Range transmission, which allowed changing speeds on the move in each of four ranges. Most important was the closed-center hydraulic system, one of the most historic and, later, most copied systems for larger tractors. The single hydraulic pump furnished power for the full hydrostatic power steering, power brakes, three independent "live" hydraulic circuits, and a 3-point hitch that used lower-link draft sensing.

Other features were similar to the 2010s, but the engines had conventional individually replaceable wet sleeves. A foot throttle provided override of the hand throttle for road use, as the Model "H" had many years previously. The 3010 engine was equipped with a balancer—standard company practice on 4-cylinder engines.

Planetary gears in the rear axles provided a smooth and durable final drive in these powerful tractors. The PTO, engaged by an independent hand clutch, could be 540 or 1000 rpm at the rear and 1000 rpm for the mid PTO. The change in speed at the rear was obtained easily by changing stub shafts.

With the unique feature of the front-mounted fuel tank, the air intakes for the coolant and oil cooling radiators were at the side. The engine air cleaner drew its supply from a small grille at the front above the fuel tank. A deluxe muffler gave both models a quiet but throaty sound.

This gasoline 3010 standard tractor (and all other standard tractors) had a shorter wheelbase than the row-crops, for a shorter turning radius. This was obtained by simply rotating the front pedestal 180 degrees to give the two different wheelbases.

A 3010 diesel row-crop tractor with adjustable front axle operates a 216-WS wire-tie baler in irrigated alfalfa.

A 3010 row-crop diesel tractor operates a 406 lister planter. From the beginning, the diesel option accounted for about 60% of 3010 sales.

A preproduction gasoline-engine 3010 tricycle tractor and PTO-driven 30 combine harvest two rows of soybeans.

The 4010—the Large-Acreage Tractor

The most popular of the 10 series tractors was the 6-cylinder 4010, which revolutionized large-scale farming. The diesel 4010 produced 84 hp at the PTO in Nebraska tractor tests. This tractor set a completely new standard with both tillage and PTO-driven machines. It was quickly accepted as the model that all other manufacturers tried to copy. The three model options were the row-crop, Hi-Crop and standard, all offered with the same three alternative fuel engines as the 3010s. Similarly, the various features mentioned for the 3010 applied also to the 4010.

All the 10 series tractors used the concept of high horsepower-to-weight ratio. This is well illustrated when the 4010 diesel is compared to the 730 diesel it superseded. A 4010 without added ballast weighs less than a 730 but produces about 37% more drawbar power. The result is increased efficiency, with a smaller proportion of power needed just to move the tractor. Higher speed operation becomes practical.

Another change in concept for the New Generation tractors was the use of variable-speed engines. Tractors could be worked in a higher gear at part throttle for greater economy on light loads. The engines were governed from 600 to 2500 rpm and worked efficiently on the drawbar from 1500 to 2200 rpm.

A single-front-wheel 4010 diesel with 2-row mounted 77 cotton stripper leaves a clean field in this once-over harvesting operation near Plainview, Texas.

A 4010 row-crop tricycle tractor with the optional LP-gas engine subsoils with a Killefer Works (Los Angeles) 23 toolbar. LP-gas engines captured over 10% of the 4010 market and outsold gasoline engines.

The only diesel 4010 row-crop tricycle tractor in the United Kingdom, No. 57638, was shown fully restored at Birdlip, Gloucester, on May 18, 1990.

This 4010 gasoline-powered row-crop with wide front end is running a PTO-driven 10A hammer mill.

A 4010 diesel Hi-Crop with 67 tool carrier and border disk increases the height of the bed.

A 4010 diesel standard operates two CC-A field cultivators. Diesel engines were especially popular on standard tractors and accounted for over 80% of sales for all types of 4010 tractors.

The 5010—the World's First 100-hp 2-Wheel-Drive Tractor

Added to the line in 1962, the huge 5010 set fresh standards for the grain farmer. With its 6-cylinder 531-cu.-in. diesel engine and new 24.5" x 32" tires, it registered 108 drawbar hp at the Nebraska tractor tests. Matched to a 7-bottom F245H plow it could outperform crawler tractors previously used, and had the added advantage of mobility.

The 5010 initiated the Category 3 3-point hitch for large mounted implements with its special combined Category 2/3 Quik-Coupler. A cab was an option available from the factory. The 5010's principal features were the same as the two smaller standard models.

The 8010—a Giant 4-Wheel-Drive Tractor

The first indication to the farming community that the company was considering a move away from the long-lived and long-loved 2-cylinder concept had occurred in 1959 at the John Deere Field Day at Marshalltown, Iowa, with the announcement of the giant 8010 4-wheel-drive tractor with 200-plus engine horsepower.

This massive 19,700-lb. machine was 8' 2" high to the top of the steering wheel, 8' wide, and 19' 7" long from stem to stern. The 8010 had a General Motors 671E 2-cycle 6-cylinder diesel engine driving a 9-forward-speed transmission with speeds from 2 to 18 mph. The tractor had articulated power steering and air brakes for easy control.

However, it was ahead of its time; the market was not ready for anything this big. One tractor was sold in 1960, and only one hundred 8010 and 8020 tractors in the seven years of sales.

To summarize, the line for 1963 consisted of six size models with 19 basic versions. Most of these offered a choice of three fuels. The best sellers were the row-crop models, with four front-axle choices. In addition there were 1010, 2010 and 3010 row-crop utility models; 3010, 4010 and 5010 standard models; 1010 utility and single-row-crop; 1010 and 2010 agricultural crawlers; 1010 and 3010 grove and orchard models; 2010 and 4010 Hi-Crops; plus the 8010 4-wheel-drive giant.

A 5010, with 27' LW disk, tills a wide swath in a large, level field. It could also pull a 34' field cultivator, a 20' tool carrier for chisel plowing or a 40' span of grain drills.

As a tribute to the late Lloyd Bellin, the owner of one of the finest collections of vintage John Deere tractors in the U.S., this photo was taken some years ago at his Isanti, Minnesota, home. Both the 8010 engine and transmission experienced some reliability problems, necessitating rebuilding 8010 tractors into 8020 tractors. This 8020 is serial number 1,078.

The 121-PTO-hp 5010 tractor brought a new level of productivity to large PTO-driven equipment. It is harvesting sorghum on a farm in Kansas with a 2-row 12 forage harvester and 115 Chuck Wagon.

An 8010 is shown at work with two 21' FW disks in 1960. It was also a match for the 37' chisel plow or four CC-A field cultivators with a 46' sweep.

Preproduction 8010 with the F180 integral 8-bottom plow shown on page 90 of Volume One, with William A. Hewitt at the controls.

The 20 Series, Improvements on a Winner

When one of John Deere's early partners questioned him for trying to improve what was already acknowledged as the best plow, Deere answered, "We must continue to improve our product or others will, and we will lose our trade." Three years after the introduction of the 10 series tractors, the 4-cylinder 3020 and 6-cylinder 4020 replaced their predecessors.

Engineers improved the original models to the extent that the 4020 became the most popular tractor of its era, and arguably one of the three or four classic tractors of all time. In the tradition of such famous tractors as the Models "D," "A" and "G," it was again another leap ahead of the competition. The 4020 became so popular that in 1966 there were 27,416 units sold in the U.S. and Canada, accounting for 48% of all John Deere tractor unit sales.

With these two new 20 series tractors came Power Shift, an 8-speed-forward, 4-speed-reverse shift-on-the-go transmission, which gave the first one-lever, no-clutching gear change under load since the All-Wheel-Drive tractor of 1918. Another useful option, introduced in 1964, was the differential lock, which provided additional traction to get through the tough spots.

A 3020 row-crop with adjustable front axle, Roll-Gard structure and 48 farm loader with grapple fork option.

The 3020 diesel standard tractor had handholds relocated on the cowl and fender. Also, the dust shield at the front corner of the platform required the addition of a second mounting step.

A 3020 diesel Hi-Crop tractor cultivates a young sugarcane crop planted on beds.

A gasoline 3020 row-crop tractor combining soybeans in October 1963 with a 42 PTO-driven combine.

A 3020 orchard tractor spraying in an orange grove with a 90 sprayer at Winter Haven, Florida, in March 1964.

A 4020 diesel standard tractor is shown with cane and rice rear tires, but without the wheat-country dust shield at the corner of the platform.

Here's the 133-hp 5020 diesel standard tractor, as sold from 1965 through 1968.

To complete the Waterloo 20 series tractors, the 5020 was announced in 1965 with its power increased from 121 to 133 hp. This most powerful standard tractor in the world became available as a row-crop model in the fall of 1966 and was supplied with 38" single or dual rear wheels in place of the standard's 32" wheels.

Later the 5020 was uprated to 141 hp and the standard model was also offered with a dual-rear-wheel option. All 5010s and 5020s had the new Category 3 3-point hitch option, with a heavy-duty Quik-Coupler to match.

The 5020, with optional factory-installed cab, and the pull-type 106 combine were popular in windrowed wheat in the prairie provinces of Canada. The 5020 had 141 hp beginning in 1969.

This 4020 diesel Hi-Crop tractor, with optional shell fenders, clearly illustrates the generous clearance provided for sugarcane and other tall crops grown on beds.

A single-front-wheel 4020 diesel row-crop tractor supplies the power for a 277 mounted cotton stripper.

The tractor that became Deere's most famous was the 4020 diesel row-crop tricycle. Between 1963 and 1972, more than 200,000 4020s were built, about double the number of 3020s, the next most widely sold John Deere tractor since 1960.

New Worldwide Tractors Unveiled

For the 1966 season, the Dubuque line benefitted from the clean-sheet design approach that the Waterloo line had experienced in 1960. The 1010 had been little more than a 4-cylinder variation of its 2-cylinder predecessor. The 2010 had many of the features of its bigger Waterloo brothers but lacked their reliability. Neither of these tractors were well matched to needs in Europe or other overseas markets. Thus, an all-new design for worldwide markets was introduced with new 3-and 4-cylinder 300 series engines. The 1020 had a 38-hp 3-cylinder engine and the 2020 a 53-hp 4-cylinder engine. Diesel or gasoline fuel options were available, but no LP-gas. The use of LP-gas was declining for large tractors and it had never been as popular for small tractors.

Big-tractor features included front-mounted fuel tank, 8-speed collar-shift transmission on both models, and the company's exclusive closed-center hydraulics with up to three independent live circuits.

This is the higher HU version of the three heights offered in the 1020 utility tractors. It provided crop clearance for cultivation comparable to the larger 3020 and 4020 tractors.

The long-legged clearance of the 1020 Hi-Crop is obtained by long spindles in front and a drop axle in the rear, similar to Waterloo Hi-Crop tractors. However, there are no added braces on the front spindles. Also, the operator is closer to the ground, straddling the transmission.

Here's the smaller 820 with F45 3-bottom integral plow in July 1969.

Here's the 1020 orchard model at work in familiar surroundings with a No. 2 integral forklift carrying a pallet of fruit. Note the special heavy front weight.

Add to this a differential lock, plus a 3-point hitch with lower-link draft sensing and their performance represented another leap ahead in the smaller tractor field. All styles had wide front ends. There were LU, RU and HU (low, regular and high utility) models having 17", 20" and 24" clearance respectively, with different rear wheel equipment and matched front axle spindles. The 3-point hitch provided a choice of load, depth, or load-and-depth control to cope with different soil conditions.

The mass market for tractors in Europe was for the size range of the 1020 and 2020, while in the U.S. and Canada, the mass market was for the 3020 and 4020 sizes. Thus, Mannheim made several 3- and 4-cylinder models. In 1968 the 1020 was bracketed in cost and power by two new 3-cylinder models for the U.S. market. The 31-hp 820 was imported from Germany and the 46-hp 1520 was made in Dubuque.

The 1520, shown in this 1968 photograph with a 640 rake in alfalfa near Linwood, New York, was popular for haymaking throughout the dairy states.

The LU version of the 2020 is shown here with the underslung exhaust. The smaller rear wheels reduced the price of this model and made it easy for the operator to mount and dismount.

Variations of the 1020, 1520 and 2020 are shown here. The 1020 has the swept-back front axle, the 1520 HU has the optional rack-and-pinion adjustable rear axle, and the 2020 has the optional air cleaner stack.

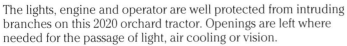

Here's an overhead view of the sleek 2020 orchard tractor. Controls were located to the rear of the rear axle for easy operation either seated or standing. This model reverted to the hand-operated transmission clutch, a desirable feature when standing.

The lights, engine and operator are well protected from intruding branches on this 2020 orchard tractor. Openings are left where needed for the passage of light, air cooling or vision.

Canada Opts for Mannheim Tractors

Canadian John Deere dealers had received their smaller tractors from the Dubuque Works since the days of the Model "M." A pilot run of the 710 tractor started a change; and with the arrival of the 20 series, the Mannheim Works in Germany was chosen as the source for utility tractors in Canada. The models selected initially were the 36-hp 920 and 45-hp 1120, to be followed a year later by the 2120. This variance in model numbers from the U.S. was to persist until the introduction of the 50 series in 1982. However, some Dubuque models were also sold at times.

A 920 tractor hurries the 3-bottom F45 integral plow along in wheat stubble as rain clouds threaten.

The 2120 tractor, with its longer wheelbase and 4-cylinder engine, was a popular loader tractor on dairy farms in the eastern provinces of Canada.

The 1120 tractor controls were typical of all 20 series utility tractors. The gearshift on the right provides four speeds, while the shift on the left provides high range, low range, reverse and park. The adjusting handle of the 3-point hitch top link also secures the link when not in use.

A 710 tractor powers a 24T hay baler as its ejector lofts a bale toward the wagon.

The 1120 tractor handles four bottoms on the F45 integral plow.

The 2510 Tractor Announced

The requirement for a tricycle tractor smaller than the 65-hp 3020 was met in late 1965 with the introduction of the 54-hp 2510. This was done in anticipation of the replacement of the 10 series utility tractors with the 20 series, which offered no tricycle option. The 2510 was built in Waterloo and was essentially a 3020 with a 2020 engine from Dubuque. The tractors had either Syncro-Range or Power Shift transmissions and were offered as row-crops with dual-wheel, Roll-O-Matic, single-wheel or adjustable-axle front end, or as a Hi-Crop model. After the 3020 was upgraded to 71 hp in 1966, a 61-hp 2520 was introduced in 1968.

The 2510 diesel tricycle operates a barely visible mid-mounted mower and a pull-type 22 crimper hay conditioner in alfalfa. Mowers were the primary use for the mid PTO, which had been standard on Waterloo row-crop tractors and was a feature of the new 20 series utility tractors.

The 2520 shows its flat, open platform and the dash-mounted single gearshift lever for Waterloo Syncro-Range transmissions. The hydraulic controls that had been on the left side of the dash on the 10 series were moved to the right of the seat for easier operation while looking back at the implement.

The 2510 diesel Hi-Crop shows greater clearance than the more widely sold 2510 gasoline row-crop with adjustable front axle. Both tractors are shown with the seat slid back for operating while standing. The seat was designed to return to a preset position when the operator sat down.

The extreme similarity of the two tractors is shown by a three-in-one "operator" advertising the gasoline 2520 tricycle and the gasoline 3020 row-crop with adjustable front axle. The 3020 is slightly taller, with a higher front grille.

Waterloo Tractors Upgraded

In 1966 the 3020 was improved and retested at Nebraska at 71 hp and the 4020 at 94 hp. Additional changes were made and the gasoline units were again tested in 1969. Engines had refinements in pistons, rings, cylinder block and liners; there was a new dry-type air cleaner, and an alternator in place of a generator. These new tractors were distinguishable by their oval mufflers. In addition the 3020 and 4020 offered a new option; power front-wheel drive, giving up to 20% boost in traction for difficult conditions. The name "Classic," gained by the first 4020, has remained with these later tractors.

A 3020 gasoline row-crop tractor, with adjustable front axle, cultivates eight 20" rows of corn with a rear-mount cultivator. The special tall, narrow rear duals were normally used in 22"-row sugarbeets. The hour meter on the front suggests this may be a test tractor.

The hydrostatic power front-wheel drive introduced during the reign of the 4020 and 3020 provided additional traction with a minimum of other changes. The adjustable front axle was offset to provide adequate crop clearance with the use of medium-size front wheels, which allowed a short turning radius.

A 3020 Hi-Crop is shown here with its front and rear axles set in a narrow position for straddling a single row of a tall or bedded crop. Hi-Crop tractors are widely used in sugarcane for growing the crop and hauling it to the sugar mill.

A 3020 diesel standard tractor cutting wheat with a 290 windrower, a common practice in the Dakotas and the prairie provinces of Canada.

The optional power front-wheel drive on this 3020 diesel tractor provides additional traction in difficult conditions.

Power front-wheel drive helps this 4020 diesel standard tractor through the muddy spots while disking.

Waterloo seriously considered introducing a deluxe cab on the 20 series tractors but decided to wait for the new styling of the 30 series to bring out the Sound-Gard body. This experimental 4020 diesel appears to have a row-crop adjustable front axle matched to the wide rear tires of a standard tractor. The last 4020 built (serial No. 270,288) was a diesel row-crop with adjustable front axle and had a conventional cab.

The 4000 tractor was the answer provided for farmers who liked the power of the 4020 but wanted the price of the 4010. This 96-hp tractor, first sold in 1969, was essentially a minimum- specification 4010 with a 4020 engine. Note the experimental modifications: the longer yellow identification stripe, adjustable step and different seat cushion.

The 4020 diesel standard was a popular tractor in the wheat growing areas of the U.S. and Canada.

The photogenic Hi-Crop appears again as a 4020 diesel with oval muffler.

This 4020 diesel row-crop tractor, with adjustable front axle, works easily with the PKM wheel-type offset disk, introduced in 1968 by the Killefer Works in California. It featured conical disk blades, which shed sticky soil better in some conditions.

Turbocharged Tractors Arrive

Late in 1968 the 122-hp turbocharged 4520 was announced, the company's first turbocharged tractor. An exhaust-driven fan forced more air into the engine's combustion chambers, allowing more diesel fuel to be adequately burned to produce more power. Although turbocharged engines are somewhat more fuel efficient, the additional power meant additional heat and thus a bigger radiator and fan. There are higher loads within the engine so the lubricating system had a major redesign. This new system had twice the capacity of the 4020s.

A "super" 4020, the 115-hp 4320, appeared in late 1970 with a Turbo-Built diesel engine. At the same time the 4520 was replaced by the 135-hp 4620 with turbocharging and intercooling, a first on a farm tractor. Cooling the incoming air makes it denser, so it's possible to squeeze even more air into the engine for even greater power output. Both models could be purchased with the power front-wheel drive option. Syncro-Range transmission was standard on both models, but the 4620 also had the Power Shift option.

The 4320 standard, with Roll-Gard structure, has the same wheelbase as the row-crops because the front pedestal is mounted the same way.

A 4320 row-crop with adjustable front axle pushes snow in the parking lot of Degelman Industries in Regina, Saskatoon. Degelman was the manufacturer of agricultural bulldozers sold by John Deere.

A 4320 tricycle tractor is shown with Roll-Gard structure, seat belt, and the SMV (slow moving vehicle) emblem behind the seat back.

A 4320 with power front-wheel drive loads an 8-row 1290 planter on a 101 implement carrier.

The 122-hp 4520 tractor gave the row-crop farmer an impressive 25% increase in available power, enough to operate the F1450 7-bottom semi-integral plow.

A 4620, with factory-installed air-conditioned cab, pulls a disk tiller, a popular tillage tool in the wheat fields of the prairie provinces of Canada.

Deere Returns to 4-Wheel-Drive Tractors

There was a 2-year absence from the 4-wheel-drive tractor market after the last 8020 was sold. Then, as an interim measure, two models of Wagner articulated 4-wheel-drive tractors were sold as engineers developed the first 4-wheel-drive tractors with field-proved John Deere components.

The 7020 was introduced in 1970 with the same basic turbocharged and intercooled 404-cu.-in. engine used in the 4620; it developed 146 hp at the PTO. Most of the parts used in the transmission and axles had already proved themselves in other large John Deere wheel tractors. The 7020 made 4-wheel-drive tractors practical for row-crop farming for the first time. Single and dual wheel combinations were available with rack-and-pinion adjustment to fit 30" and 38" rows.

The 7520 was an immediate success when first sold in 1972. Over 600 7020s had been sold in the U.S. and Canada in 1971. In 1974, over 2,000 7520s were sold in the same market, which was more than double the sales that year of the successful 6030. The 7520 provided the additional power that customers were requesting by using the 531-cu.-in. engine developed for the 6030, which produced 175 hp at the PTO in both models.

A 7520 with a gang of six press-wheel grain drills crosses a Western plain. The 7020, its twin with a smaller engine, is shown on page 31.

The WA-14, built by Wagner, had only limited styling changes made for John Deere. It produced 178 hp at the drawbar from a 225-hp engine.

The Wagner-built WA-17 produced 220 hp at the drawbar from its 280-hp engine. Combined sales of the two models for the three years they were available were approximately 60 units.

The 30 Series, Generation II

The first 30 series tractor to appear was the 2030, which replaced the 54-hp 2020 in the fall of 1971. It continued the trend of more horsepower in a given size package. In 1971 Nebraska tractor tests, both the gasoline and diesel versions produced 60 hp. The 2030 was the last gasoline tractor produced by John Deere to be tested at Nebraska. Gasoline versions accounted for about one-fourth of all 2030 sales when production of that option was stopped in 1973.

The 2030 had a constant-mesh transmission with Hi-Lo or direction-reverser options. It continued to have all the other time-proven features of the earlier model.

In 1973, the 3-cylinder 31-hp 820 from Mannheim was upgraded to the 35-hp 830. The 3-cylinder 1520, which had been built at the Dubuque Works, was replaced by the 1530, built at the Mannheim Works, with a slight reduction in power (from 46 to 45 hp). Dubuque continued the 4-cylinder 60-hp 2030 and added the completely new 4-cylinder 70-hp 2630. All four models retained the earlier styling and were promoted as low-priced tractors in comparison to the competition.

The diesel-powered 2030 tractor with integral 100 disk is a maneuverable combination for small fields. A 2030 gasoline tractor is shown on page 34.

The rotary fuel injection pump, shown below the muffler on this 2630 tractor, is the type used on all 300 series diesel engines.

The 1530 tractor, shown with ROPS but no canopy, is almost identical in appearance to the 830. The 1530 was normally sold with slightly larger tires than the 830.

The lowest cost combination for row-crop cultivation was the 830 tractor with RR2 cultivator shown here. Note the row guide behind the front axle and the stabilizing coulter to keep the cultivator on the row.

A 2630 tractor with 450 drawn mower is shown making a tight corner while clipping a pasture.

30 Series Tractors
Delayed in Canada

In Canada, as in Europe, the 920 and 1120 stayed in the line until 1975. To avoid confusion with the Dubuque 2030, the European 4-cylinder equivalent was designated as the 1830 tractor. The 66-hp 2130 was the top-of-the-line tractor from Mannheim for the Canadian market, their equivalent to Dubuque's 2630, but the former had a 239-cu.-in. turbocharged engine while the latter was naturally aspirated with 276-cu.-in. displacement. The delayed introduction of the 30 series in Canada resulted in different styling within the series.

By 1975 the smaller models in Canada had adopted the styling of the Waterloo tractors, as seen in this 1030 tractor with 2-post Roll-Gard structure.

This drawing shows an unusual combination of an underslung muffler on a regular 1830 farm tractor with Roll-Gard structure.

The 3-cylinder 1630 was a handy tractor, either as the main tractor on dairy farms or as a second tractor on grain farms.

The orchard version of the 1630 is readily recognized by its low overall height, underslung muffler, narrow tread and wide 18.4-16.1 rear tires.

The 66-hp 2130 makes a good haying teammate for the 500 round baler. In 1975, the 2130 was the largest of the Mannheim utility line and was the best seller of all 30 series tractors in Canada.

The 80-hp 3130 tractor provided the Canadian farmer with 6-cylinder power in a smaller package than available in the Waterloo models. Note the triangular "slow moving vehicle" emblem to the rear of the seat, and the reflectors mounted on the Roll-Gard posts.

Generation II Tractors—the Sound-Gard Body

John Deere announced a completely new and restyled series of row-crop tractors in August 1972, giving it a market advantage rivaling that which the New Generation had achieved in 1960. The Sound-Gard body was a revolutionary new safety cab with built-in rollover protection. It had the option of a pressurizer, to keep dust and dirt out, or a heater and air conditioner. Its large curved windshield with twin wipers gave it a unique appearance, and all the glass was tinted and polarized.

All models could have the new cab or a 4-post ROPS. In the first year of availability, the Sound-Gard body was purchased on over half of the Waterloo 30 series tractors. Adoption of the 4-post Roll-Gard structure also increased, resulting in three-fourths of the Waterloo 30 series tractors being sold with rollover protection. Additional benefits for the operator included underslung pedals, adjustable tilt-telescope steering wheel, and seat belts. The low noise level in the cab made the stereo radio with cassette player an attractive option.

Four models were introduced—the 80-hp 4030, 100-hp 4230, 125-hp 4430 and the 150-hp 4630. The 4-cylinder 3020 was replaced by the

The new 4-post Roll-Gard structure is shown on the 6-cylinder 80-hp 4030 tractor.

Sales of tricycle tractors declined along with use of mounted corn pickers and front-mounted cultivators. Tricycle tractors were also limited by the tire sizes that would fit under the tractor.

Since 1947, exclusive Roll-O-Matic "knee-action" front wheels have been making riding easier for tricycle-tractor owners. This feature automatically cuts front-end "bounce" in half and provides sure-footed stability over uneven terrain. It's a feature carried over on new 4030 and 4230 Sound-Idea Tractors.

INDUSTRY TRACTOR SALES

Annual industry farm tractor sales in the U.S. and Canada varied between 150,000 and 225,000 units for the period of 1956-1981. Canada typically accounted for 12-14% of these sales. Sales had been double this amount during 1947-1953 as many farmers bought their first tractor. Average power of tractors sold increased from 45 hp in 1960 to 90 hp in 1975. As power increased, unit sales should have declined, but high farm exports and resulting high farm prices caused 1973 and 1979 to set tractor sales records for the two decades. With the downturn in the farm economy of the '80s, annual industry sales were between 120,000 and 150,000 for 1982-1990. The economic impact of this reduction in unit sales was somewhat worse for the industry because as many as 60,000 of these units were below 40 hp and were not sold for farming.

The 2030 utility, 4630 row-crop and 7020 4-wheel-drive tractors for sale in 1972 illustrate the transitional period.

The 100-hp 4230 had adequate power to operate the 300 corn husker with 244 corn head while towing a large gravity-box wagon.

4030 with a 6-cylinder 300 series engine. The 4230 was an upgrade of the popular 4020. However, it was the turbocharged 4430 that was the best seller in the 30 series. The largest model, the 4630, had an intercooled as well as turbocharged engine.

All four models had a new hydraulically controlled Perma-Clutch that used circulation of oil to dissipate heat, providing a very long clutch life. The standard transmission for all the models was the well-proven Syncro-Range, but options for the 4030, 4230 and 4430 included a new 16-speed Quad-Range, which was a blending of the Syncro-Range and a built-in Hi-Lo no-clutch shift. The three larger models were also available with Power Shift, similar to the models they replaced.

JOHN DEERE TRACTOR SALES IN 1975

Sales are given for 1975, the midpoint of the period covered by this book. The number of units sold is typical of the '60s and '70s and the sizes are typical of those sold for farming in the '70s and '80s.

Utility Tractors, U.S.

Model	Units
830 and 2040	3,706
1530 and 2240	3,050
2030 and 2440	3,919
2630 and 2640	4,463

Utility Tractors, Canada

Model	Units
920	188
1120	593
1830	866
2130	1,903

Waterloo Tractors, U.S. and Canada

Model	Units
4030	2,562
4230	5,108
4430	13,782
4630	6,748
6030	506
8430	1,336
8630	1,822
Total Tractors	50,552

A low-profile 4230 tractor uses two stacks of front-end weights to balance a 3-bottom 2-way plow. The tractor has the short wheelbase of a standard tractor and the reduced overall height and underslung muffler of an orchard tractor.

The 4230 shown is a very rare combination of a Hi-Crop tractor with a gasoline engine.

A 4430 tractor transports an integral 1100 cultivator with harrow attachment. The 125-hp 4430 took over as the best-selling John Deere tractor in 1973, an honor that had been held by the 84-hp 4010, first sold in 1960, and the 91-hp 4020 that replaced it in 1963. Then increases in power slowed as the best seller became the 130-hp 4440 in 1978, the 140-hp 4450 in 1983, and the 140-hp 4455 in 1989.

A 4630 completes fall tillage with a 350 Level-Action offset disk soon after the corn harvest.

In rolling countryside a 4630 tractor, with optional hydrostatic front-wheel drive, prepares a field with a 100 series chisel plow.

"Muscle" Tractors of the 30 Series

The 141-hp 5020 was replaced in 1972 by the 175-hp 6030 using the new turbocharged and intercooled 531-cu.-in. engine. The previous engine, which provided 141 PTO horsepower, was offered as an option the following year for the 6030, making it the only John Deere tractor to ever provide a choice of engine sizes.

In late 1974 the company announced the 175-hp 8430 and the 225-hp 8630 4-wheel-drive tractors. The increase in power over the 7020 and 7520 came from increased engine displacement. The 215-hp engine was increased from 404 to 466 cu. in. by enlarging the bore from 4.25 to 4.563 inches. On the 275-hp engine, displacement was increased from 531 to 619 cu. in. by enlarging the bore from 4.75 to 5.125 inches.

The '70s were a time of transition for John Deere tractors. The 4-wheel-drive tractor market was entered successfully with field-proven Deere components. Turbocharged engines became the norm in tractors above 100 hp. In the final years of the 4020, LP-gas and gasoline versions each accounted for about 1% of sales. LP-gas was not offered in the 30 series and only a limited number of gasoline 4030s and 4230s were made during the first year of production. Anticipated sales of these gasoline tractors had been so low that they were not even tested at Nebraska. The upper attaching points on the tractor for the front-mounted cultivator braces were dropped after more than three decades of use. Single, dual and Roll-O-Matic front-wheel options of tricycle tractors had their final years on the smaller 30 series Waterloo tractors.

In early 1975, there were 11 sizes of John Deere 30 series tractors in the U.S. Utility tractors spanned the range of 35 to 70 hp, with Mannheim as the source of the two 3-cylinder models and Dubuque the source of the two 4-cylinder models. Waterloo provided four row-crop tractors from 80 to 150 hp, the 175-hp 6030 standard tractor, and the 175-hp 8430 and 225-hp 8630 4-wheel-drive models.

The differences between the 141-hp and the 175-hp 6030 tractors are well illustrated in these two tractors with factory-installed, air-conditioned cabs. The smaller naturally aspirated engine, shown in the background, has a smaller forward air-cleaner stack. The turbocharged engine shows its intercooled intake manifold.

An 8430 4-wheel-drive tractor is on its way to the field for a plowing demonstration at the 1974 Farm Progress Show. Manufacturers use this annual farm show, with its huge gathering of farmers, to announce many of their new products

A typical use for the 175-hp 6030 standard tractor was pulling a 6-gang HZ press drill.

The 8430 4-wheel-drive tractor and 780 disk air drill, tilling and seeding a wide swath in a large field.

This 6030 tractor is well prepared to work in muddy conditions with its dual rear cane and rice tires.

The 40 Series, More Powerful Utility Tractors

During the summer of 1975 the utility line in the U.S. became the 40 series. They had the Generation II styling and a new Roll-Gard roll-over protective structure, and were announced in even 10-hp steps. The 3-cylinder Mannheim-built tractors were the 40-hp 2040 and 50-hp 2240. Each had 5 hp more than the tractor it replaced—the 830 and 1530, respectively. Dubuque was the source for the 4-cylinder 60-hp 2440 and the 70-hp 2640. The Mannheim-built 6-cylinder 80-hp 2840 joined the line in early 1976.

The utility tractors were regularly equipped with power steering, hydraulic brakes, "live" PTO, 3-point hitch with lower-link sensing, differential lock that could be engaged on the move, planetary final drives, and a fully adjustable swinging drawbar; all features they shared with the larger row-crop models. This set them apart from competition and enabled them to be introduced with the slogan, "Family styling that's inherited—family reputation that's earned."

The Roll-Gard canopy on this 2040 tractor features the appearance and lighting of the Sound-Gard body.

The 2240 vineyard tractor has narrower tires than the 2240 orchard model. They have a similar narrow tread and low overall height.

This 50-hp 2240 tractor features full crop clearance with its rack-and-pinion adjustable rear wheels.

This later model 2040, with a 105 integral disk, has the new mechanical front-wheel drive and shows the new styling.

This 2040 tractor is working in a hayfield with a 670 drawn rake.

A 2240 with mechanical front-wheel drive is spraying a well-pruned vineyard.

The orchard version of the 2240, with a mounted sprayer, illustrates how its low, wide tires work well among trees.

For the 1980 season Deere introduced an updated series of tractors offering "New Profiles of Performance" in the 40- to 80-hp sizes, without changing any of the model numbers except on the 80-hp tractor, which became the 2940. New styling, with the hood raised in the area of the new dash, prepared these tractors to be fitted with Sound-Gard bodies.

The 2040 engine was increased from 164- to 179-cu.-in. displacement. The top-shaft-synchronized (TSS) transmission was added as an option for the Mannheim-built 2040 and 2240. Hi-Lo was an option for the 2240, 2440 and 2640; the alternative option was the hydraulic direction reverser for quick direction changes. The 2440 and 2640 had more than a 20% increase in 3-point hitch lift capacity.

The 2840 was replaced by the 2940, with a stronger engine block and displacement increased from 329 to 359 cu.in. The 2940 had the TSS transmission with Hi-Lo shift, providing 16 forward speeds. The Mannheim-built 2040, 2240 and 2940 each offered mechanical front-wheel drive with the differential offset and the tie rod in front of the axle.

The five models had more comfortable seating, an electronic instrument panel, and a new parking brake. The engine and transmission were warranted for 1500 hours or two years on all models. The new tractors represented a further advance in the medium-size tractor field.

The roll-over protective structure on the Dubuque-built 2440 provides a good framework for mounting the weather shield.

An 80-hp 2840 is shown cultivating eight rows of corn with a front-mounted cultivator. The 2840 was the first Mannheim-built 6-cylinder tractor sold in the U.S. It and its 2940 successor produced the same power as the original 4010 and were the top sellers among the utility tractors.

The 70-hp 2640 was the most popular utility model for the first three years of sales. It provided power similar to the popular 3020 of a decade earlier.

This 2440 is cutting grass and weeds with a drawn 1008 rotary cutter.

The 2940 MFWD tractor with 709 drawn rotary cutter makes a good team for cutting weeds and clearing brush.

Canadian Market Gets 40 Series Update

The 30 series Mannheim tractors marketed in Canada had the intermediate styling for utility tractors, so the 40 series were not introduced until the higher hood and new dash were available. Some of the new Canadian models no longer used the model numbers of their European counterparts. The German-sourced tractors included the 44-hp 1040 and the 50-hp 1140 orchard and vineyard 3-cylinder models; the 55-hp 1640, the 60-hp 1840 and the turbo-charged 70-hp 2140 4-cylinder models; and the 80-hp 3140 6-cylinder model.

These 1980 tractors were labeled the "Schedule Masters," since they were designed to do more in a given time. New engine features, Hi-Lo transmission option from the smallest model up, and redesigned wet-disk brakes were among the new features of the 40 series.

More significant was the mechanical front-wheel drive option on the 1640 and larger models, a development that would have far-reaching repercussions. This provided 4-wheel braking, not available on previous tractors with hydrostatic front-wheel drive.

With 53" hood height and 59" width, this 1140 orchard/vineyard tractor is equipped with vineyard size 14.9-24 rear tires. It is shown here with a rotary cutter in an orchard.

The SG2 Sound-Gard body became available in 1981 on the 2140 as well as on the 3140 tractor shown here. The SG2 cab was built in the new Bruchsal Works in Germany.

The mechanical front-wheel drive on this 1840 tractor illustrates the off-center drive and front tie rod.

A Canadian 1840 tractor, with 115 drawn disk, differs from its U.S. counterpart, the 2440, by having its muffler under the hood.

A representative lineup of power available to Canadian farmers included the 27-hp 950, 44-hp 1040, 180-hp 4840 and 228-hp 8640.

Yanmar Extends the Tractor Line

Late in 1977, after prolonged negotiations, the company announced the first two diesel tractors manufactured by Yanmar in Japan, the 3-cylinder 22-hp 850 and 27-hp 950. They were billed as "Little Big Tractors" because, despite their size, they had most of the features of their larger brothers.

The 2-lever-control transmission provided eight forward speeds, ranging from 1 to 12 mph, and two reverse speeds. Their specifications included engage-on-the-go differential lock, a 540-rpm PTO with an overrunning clutch, and a Category 1 3-point hitch. Both models had MFWD as an option. The Yanmar-built line was extended in late 1979 with another 3-cylinder model, the 33-hp 1050, followed in 1981 by the smaller 650 and 750, which are covered under consumer products.

A very full line of matched equipment was provided including two integral plows, two disks, toolbars and planters, a cultivator, rotary tiller, rear blade, mowers, a rotary cutter, post-hole digger, scraper, loader and backhoe. The Yanmar tractors had an agricultural origin in Japan. It was thought that the 850-1050 would be used with the broad line of tools for farming. They did sell well, but primarily to the large-acreage homeowner for cutting grass with a mid-mounted rotary mower, an integral grooming mower or an integral rotary cutter. Loaders and rear blades were also popular.

A 950 tractor with 2-post Roll-Gard structure is working with a 105 integral disk. Note the small suitcase weights.

The compact 33-hp 1050 tractor is shown in a hayfield. It had the power of the 520 and 530 tractors of the '50s.

Showing its versatility, this mechanical front-wheel drive 850 tractor with 75 loader fills a spreader and then cleans up the yard with a rear blade.

In 1982 another model appeared, the 143-cu.-in. 3-cylinder 1250 rated at 40 hp. To complete the series in 1984, the 50-hp 1450 and 60-hp 1650 were announced. These latter two tractors had 4-cylinder engines of 190 cu.in., naturally aspirated in the 1450, turbocharged in the 1650. The three new models set nine records when tested at Nebraska. The 1650 was the most fuel-efficient tractor tested of any size until this record was taken from it by the 4955 five years later. They all had 3-range sliding-gear, collar-shift transmissions, providing nine forward and three reverse speeds.

These three Yanmar tractors had the horsepower of the 2040-2440, which had been replaced by the 2155-2555 with five additional hp each. This provided the farmer with a choice of three 45- to 65-hp tractors from Germany and three "lean and trim" 40- to 60-hp tractors from Japan. Although the Yanmar tractors proved to be very durable in farm use, customers showed a distinct preference for the heavier, more familiar tractors from Mannheim.

A 1250 tractor cuts weeds and grass with an integral 506 rotary cutter.

Proving its adaptability, the 1250 MFWD tractor makes hay with a 327 twine-tie baler.

A 1450 MFWD tractor is hard at work chisel-plowing in a young vineyard.

An excellent use for the economical 1450 tractor is windrowing hay with an integral 650 rake.

A 1650 tractor keeps this 430 round baler busy at haying time.

A popular job for a 1650 tractor is hauling a load of fruit from a grove.

Enter the Iron Horses

Announced late in 1977, a new line of five 90-to 180-hp tractors were introduced from Waterloo. With the exception of the 4040, all the tractors had new engines of 466-cu.-in. displacement. The 4040 had a naturally aspirated 404-cu.-in. engine developing 90 PTO hp. The 4240 naturally aspirated version turned out 110 hp; the 4440 turbocharged unit delivered 130 hp; the 4640, being both turbocharged and intercooled, had 155 hp. In addition to the row-crop models, the 4240 and 4440 were available as Hi-Crop tractors with a standard 4-post Roll-Gard structure.

Farmers had asked for a Sound-Gard body on the 6030, so the new top-of-the-line 180-hp 4840, with Power Shift transmission as standard equipment, provided the solution. Using the same engine as the 8430, it provided the customer with two choices at this power level.

The 16-speed Quad-Range transmission was standard on the four other models. The new engines had bigger radiators and water pumps, larger fans and higher capacity alternators on all models. There were new fuel injection pumps, distributor-type on the two smaller tractors and in-line on the three larger. Bigger fuel tanks met the farmer request to run 10 hours without refueling. Hydraulic capacity was greatly increased for the 3-point hitch, for remote cylinders and for loaders.

The entire 40 series of tractors from Waterloo gave "more horses from more iron," enabling them to truly live up to their name: the Iron Horses. To emphasize the advances made, a new 2-year warranty was introduced on these engines and the heavier powertrains.

The smallest of the Iron Horses, the 90-hp 4040, is shown with a Sound-Gard body.

Decals can be deceptive when the equipment is owned by product engineering departments. The decals suggest a 307 loader on a 4040 tractor. However, it is an experimental 260 loader on an experimental 4050 tractor with mechanical front-wheel drive. This is an example of how implement factories work with tractor factories so new implements are available when new tractors are ready for sale. Note also that the muffler is under the hood.

The best-selling 40 series tractor was the 130-hp 4440. It's shown transporting a 1508 rotary cutter.

For the frequently muddy harvesting conditions, this 4640 has power front-wheel drive and narrow rear duals that straddle sugarbeet rows about to be harvested with the 4310A beet harvester.

A new 4640 tractor equipped for dual rear wheels is loaded on a dealer's trailer at the Waterloo Tractor Works.

Soft conditions like this make good use of both the power front-wheel drive and the rear duals on this 4440 tractor.

The company introduced two new 4-wheel-drive tractors in 1979, with slightly increased horsepower over the 8430 and 8630 they replaced. The 8440 and 8640 were the result, with optional front and rear hydraulic differential locks, four remote cylinder outlets, a new single-lever Quad-Range transmission control, and a new HydraCushioned seat for increased operator comfort. Add to this an Investigator warning system to monitor many engine and power train functions for the extra hours expected from operators of these large machines.

In 1981, John Deere was offering the broadest range of tractor sizes and types in their history. The U.S. customer could choose among three compact utility tractors with 22 to 33 hp from Japan, five utility tractors with 40 to 80 hp from Germany and Dubuque, five row-crop models with 90 to 180 hp from Waterloo, and two 4-wheel-drive models with 180 to 228 hp from Waterloo.

An 8640 tractor with multiple hitch pulls four 9350 hoe drills with 4.00x18 pneumatic press wheels.

TRACTOR FIELD TESTS

New tractors and new features for tractors are continually being developed and tested in the field. At the time the 40 series tractors were being sold, the 50 series were being developed. The goal is to develop the functional and productivity performance the customer needs, along with good durability and reasonable cost. In 1981, the Waterloo Product Engineering Center had experimental components on more than 1,000 tractors distributed throughout the U.S. and in Australia, England, Germany, Malaysia and Mexico. Competitive tractors also are tested and evaluated in the field and in the laboratory.

A decade ago there were field test sites with good shops, instrumentation and trained technicians in Arizona, California, Colorado and Texas. Sites are chosen to include a variety of crops, soils, and climate. Preference is given to large farms that can provide a long work season and good security from outside observers. More than 60,000 hours of useful farm work was done with these test tractors. A smaller number of non-productive hours were accumulated, just to get more hours of durability testing. The average test tractor gets three to four times as many hours of use per year as the average tractor in the Corn Belt.

The implement factories cooperate in functional and durability testing of their new equipment with PEC at Waterloo and the remote test sites. Implements are generally tested in more states in a greater variety of crops, soils and moisture conditions than tractors.

A used 8640 trade-in tractor in excellent condition is shown in an Illinois dealer's yard on September 23, 1989.

Like their 2-wheel-drive brothers, the 4-wheel-drive 8440 and 8640 tractors had many improvements for function and durability but little change in appearance from their predecessors.

The largest of the Iron Horses, the 180-hp 4840 tractor, is shown operating a 770 air disk drill.

The 50 Series, with Caster/Action MFWD

Despite the adverse market conditions late in 1982, the longest new line of John Deere tractors ever introduced by the company was announced. Eleven models from 45 to 190 PTO hp included five utility models, an orchard/vineyard model and five row-crop models. The Dubuque Works concentrated their production on industrial equipment, and Mannheim became the single source for utility tractors in the U.S. and Canada. With this change, the model numbers in the two countries became the same.

The utility models built in Mannheim were the 3-cylinder 45-hp 2150; 4-cylinder 55-hp 2350, 65-hp 2550 and turbocharged 75-hp 2750; and 6-cylinder 85-hp 2950. In addition a 3-cylinder 50-hp orchard/vineyard model was designated 2255, coming with an engine size between the 2150 and 2350.

The new Efficiency Experts were included in a November 1982 advertisement.

The 2255 vineyard model used 14.5-24 rear tires to minimize tractor width between the rows of grapevines.

The 2150 had adequate power to operate a 330 round baler, shown discharging a picture-perfect round bale of alfalfa.

New Caster/Action mechanical front-wheel drive was available on all 2350 through 4850 tractors. Few customers had chosen MFWD on the 3-cylinder models, so the previous design with front tie rod and offset differential was kept to minimize cost. The Sound-Gard body was a new option for 4-cylinder tractors but not offered on the 2150 because of the added cost, interference with loaders and top-heavy appearance.

The orchard version 2255, shown disking between trees, featured 18.4-16.1 rear tires for good flotation and minimum soil compaction.

They were introduced as The Efficiency Experts, a timely designation because fuel represented 35% of farm tractor costs in the early '80s compared with only 25% in 1970. The new 8-speed top-shaft-synchronized (TSS) transmission offered Hi-Lo or direction reverser as options on these new tractors. Each model had 5 more hp. Servicing was simplified, with all the daily checkpoints located on the right side or rear of the tractor.

The independent PTO on the 2550 and all other utility tractors made the stop-and-go operation of the 430 round baler an easy task.

The base 2350 tractor had 2-wheel drive, Roll-Gard frame and flanged rear axles. It is shown with a chisel plow in a vineyard.

Mechanical front-wheel drive gives this 2550 tractor with 245 loader added traction in a muddy beef cattle feedlot.

The Sound-Gard body on the 2350 tractor protects the operator from dust and heat as he bales with a 430 round baler. The rack-and-pinion rear axles simplify tread adjustments for row-crop cultivation.

Mechanical front-wheel drive greatly improves mobility in muddy conditions for this 2350 tractor with 245 loader and pallet fork.

The 95-hp 3150, introduced in 1985, was the first John Deere tractor with mechanical front-wheel drive as standard equipment. The MFWD on this model engaged automatically when additional drawbar pull was required, but the operator also could engage it under normal power or traction demands. The 3150 had a 16-speed TSS Hi-Lo transmission and a naturally aspirated 359-cu.-in. 6-cylinder engine. With a 3-point hitch lift capacity of 6300 lb, it could handle wide equipment comfortably. With the same power as the classic 4020, plus many new features, the 3150 was indeed a worthy successor to that most popular of tractors. A picture of the 3150 is shown on page 62.

Several specialty models were added to the line in 1985, including a low-profile 2750; a high-clearance 2750 Mudder tractor with MFWD; and four wide-tread models, the 2350, 2550, 2750 and 2950, for use in wide-row crops like tobacco and some vegetables.

This 2550 tractor with MFWD is equipped with an underslung exhaust muffler for orchard spraying.

The Sound-Gard body had excellent visibility for plowing with MFWD, as shown on this 2550.

A 2750 tractor is shown with an 8000 series end-wheel drill in a well-tilled field.

This low-profile 2750 is operating an integral chisel plow in an orchard.

A 2750 Mudder tractor being transported by the Fresno, California, dealer on September 26, 1986. The large-diameter tires provided high crop clearance and the MFWD improved both steering and traction in marginal field conditions.

The following year another specialist tractor was announced, the 25-hp 900 HC from Yanmar. Designed especially for growers who needed very accurate control when cultivating a single row of tender young plants in nurseries and tobacco fields, the tractor was offset—reviving memories of the Models "L" and "LA" of the '30s and '40s.

High clearance made the 900 HC tractor ideal for cultivating tobacco and vegetable crops. The offset positions of the operator and the engine provided a clear view of the crop.

The Sound-Gard body on this 2950 tractor with MFWD keeps the operator warm and dry as he clears the farmstead with a 686 snow blower.

Here's a 2950 tractor and 530 round baler putting up alfalfa hay. A 2550 tractor and 430 round baler are in the background.

The 1-row 25-hp 900 HC tractor was a good match for many North Carolina farmers with small tobacco acreage allotments.

Waterloo Offers 15-Speed Power Shift

The Waterloo models in the 1982 announcement were the 100-hp 4050, 120-hp 4250, 140-hp 4450, 165-hp 4650 and the 190-hp 4850. All of these tractors had 6-cylinder engines and were available with a new 15-speed Power Shift transmission (standard on the 4850).

John Deere's new Caster/Action mechanical front-wheel drive, with 13 degrees of caster, allowed these tractors to have a short turning radius, even when wheels were set for 30-inch row cultivation. Customer adoption of MFWD was much higher than for the preceding hydro-static front-wheel drive, because of added traction from the positive drive and larger front tires. The two larger models with MFWD now had enough drawbar pull to be more competitive with 4-wheel-drive tractors.

Optional fenders are shown on this 4050 tractor equipped with the new Caster/Action mechanical front-wheel drive.

This 4250 tractor illustrates the centered differential and the rear tie rod of the Caster/Action mechanical front-wheel drive. The axle pivot is located above the differential to eliminate tire contact with the frame during extreme turns combined with maximum axle oscillation.

The Sound-Gard body makes this 4250 Hi-Crop a safer and more comfortable tractor, for example in Louisiana sugarcane fields.

The air-conditioned comfort of the Sound-Gard body on this 4050 tractor was fully appreciated when used with a 250 drawn sprayer.

A 4250 clips a pasture with an integral 1408 rotary cutter late in 1982.

Here's a 4450 tractor with a drawn chisel plow and tine harrow attachment. The 4430 and 4440 had been best-selling John Deere tractors. In the 50 series, the 4450 was the most popular. It accounted for almost one-fifth of the sales of the 17 sizes of tractors.

The row-crop models represented a further advance with their updated Investigator II electronic warning system, new selective hydraulic control valves and new breakaway hydraulic couplers. A Quik-Coupler hitch was standard on the 4650 and 4850 tractors. Various records were again set at the Nebraska tests, including one for the quietest tractor ever tested there. The 4050 measured 70.0 dB(A) in the Sound-Gard body at 50% load and 73.5 dB(A) at full load.

Two Hi-Crop models, the 4050 and 4250, were offered at first; both had 41 inches of under-tractor clearance and 44.6 inches between the final drives. However, demand was for the larger model, so the 4050 Hi-Crop was dropped from the line.

This 4450 tractor with MFWD takes on a different appearance when surrounded by a 740 cotton stripper. Note how the front fenders have been reversed to protect the unharvested cotton.

①

②

③

A pair of 4450 tractors with Caster/Action MFWD illustrate how the front wheels lean to provide shorter turns without frame interference. Note the step on the front axle and the handhold for easier refueling. Optional front fenders are shown on the left tractor.

① This field-ready 4650 tractor has dual rear wheels and a full complement of 20 front suitcase weights. The wider fuel tank and radiator in the 4650 and 4850 tractors permitted the use of five front headlamps, while the 4050 through 4450 tractors had only four. All Waterloo 2-wheel-drive tractors continued with the 20-inch-wide frame, introduced in 1960 on the 3010 and 4010 tractors.

② A rear view of the same 4650 tractor illustrates the challenge of providing a spacious cab with the 60" tread required to cultivate 30" rows. The Quik-Coupler could be left on when using the drawbar. The tractor has an SMV emblem, reflectors, taillights, and four round rear work lights.

③ A 4650 tractor is shown with a 712 mulch tiller in corn stubble. The sides of 4650 and 4850 tractors are indented to provide turning clearance on mechanical front-wheel drive models.

This 4850 tractor with MFWD easily handles two drawn 750 no-till drills.

A 4850 tractor with 722 mulch finisher and sprayer completes soil preparation prior to planting.

A V8 Added as the Top 4-Wheel-Drive Tractor

In 1982 the 4-wheel-drive 8440 and 8640 tractors were replaced by the 185-PTO-hp 8450 and 235-PTO-hp 8650. John Deere also announced their largest tractor ever, the 300-PTO-hp 8850. It featured a turbocharged and intercooled V8 diesel engine designed and built by John Deere at Waterloo.

These new 4WD tractors were produced in response to farmers' continuing request for more power. Advertised as "Three new ways to tighten your belt" in the increasingly difficult conditions of the day, they offered increased power, increased productivity and reduced operating costs. Servicing was made easier by a grouping of all the service points and by tilt-up hoods, providing better access.

All three models had improved operator visibility, with the muffler and air cleaner intake moved near the right front corner of the Sound-Gard body. New ISO remote hydraulic couplers (up to four) and a Category 4/4N Quik-Coupler hitch on the 8850 ensured maximum adaptability with both drawn and mounted implements.

The monitor system was updated to cover more functions and was renamed Investigator II. New and brighter front lights, relocated at the top of the grille, improved visibility.

Single-lever Quad-Range transmission with 16 speeds and a hydraulically engaged Perma-Clutch were standard on all three models. Economy was further enhanced with the use of mixed-flow fans on the 8450 and 8650 and a temperature-sensing viscous fan drive on the 8850. The latter reduced fan speed automatically when the temperature dropped.

This 8450 4-wheel-drive tractor is shown with an 845 folding row-crop cultivator in soybeans. The wheel tread of these 4WD tractors could be adjusted for 30" or 38" rows. However, most customers preferred the steering of traditional 2-wheel-drive tractors for cultivation.

This 8850 with 712 mulch tiller completes fall tillage soon after the corn is combined.

The 8650, 8850 and 8450 family of 4-wheel-drive tractors show the new lights, offset muffler and air cleaner intake position. The wider grille of the 8850 has six headlamps, compared with four on the 6-cylinder models.

The 8650 tractor easily powers the PTO-driven 7721 combine in windrowed grain. In work like this, single wheels are used as the ground is dry, the drawbar load is light and the narrower width is desirable.

The 8850 tractor is fully loaded with an integral 910 V-ripper operating deep in hard post-harvest soil conditions.

The 55 Series, Refinements on Proven Designs

The next new tractors were introduced in 1987, the sesquicentennial anniversary of the company. The Mannheim-built line was refined and made more productive. The new models were the 3-cylinder 45-hp 2155 and 55-hp turbocharged 2355N orchard/vineyard models; 4-cylinder naturally aspirated 55-hp 2355 and 65-hp 2555 (with standard transmission); 4-cylinder turbocharged 65-hp 2555 with top-shaft-synchronized (TSS) transmission, 75-hp 2755 and 80-hp 2855N; and 6-cylinder 85-hp 2955.

With 2-wheel drive or MFWD, and a transmission choice of collar shift, TSS, TSS with Hi-Lo providing 16 speeds, TSS creeper with 12 speeds, or an 8-speed collar shift with hydraulic direction reverser, the customer could customize a tractor to suit his needs. When the optional TSS transmission was chosen, rated engine speed was reduced from 2500 to 2300 rpm.

Among the specialty models introduced in 1987 and 1988 were two orchard/vineyard models, and the 2755 and 2955 high-clearance models (previously called Mudders). An optional 96" front axle made the 2355 through 2955 into wide-track models. For the non-farm market, there were the four general purpose models, the 2155 through 2755, usually supplied in highway yellow.

In 1988 the 95-hp 3155 MFWD replaced the 3150 MFWD, completing the Mannheim-built 55 series tractor line. Like the 3150, it had MFWD as standard equipment.

This 2155 GP tractor, available through consumer products dealers, features wide tires with shallow tread to minimize turf damage.

A 2355, with underslung exhaust and flanged rear axles, is shown with a 450 drill at Klug Farm Equipment, Eau Claire, Michigan, in March 1991. Dealers started using this distinctive and attractive sign in 1968.

This 2155 tractor shows its versatility by operating a 350 sicklebar mower with a 175 loader attached. Caster/Action mechanical front-wheel drive replaced the previous MFWD.

This well-equipped 2355 has both Caster/Action mechanical front-wheel drive and a Sound-Gard body, a good combination for the Northern dairy farmer. The MFWD helps in loader operation and the Sound-Gard body provides all-season comfort and protection for the operator.

Here's a 2355 tractor and 327 baler handling a heavy cutting of alfalfa in the Midwest.

This 2355 general purpose model, sold by an industrial equipment dealer in Florida, features a grille guard and high-flotation tires for sandy terrain.

This 2355N tractor is shown with an orchard sprayer for spraying hops. The non-Deere cab has a sloping front to minimize damage to the hop vines.

Here's a 2555 MFWD tractor baling alfalfa with a 335 round baler.

A basic 2555 tractor is shown in front of a farm equipment dealership, Goodrich Equipment Co., Geneseo, Illinois.

This 65-hp 2555 tractor shows the design of the Caster/Action mechanical front-wheel drive with fenders. Notice the clear view allowed by the Sound-Gard body.

JOHN DEERE DESIGN, DEPENDABILITY, AND DEALERS MAKE THE DIFFERENCE

This advertising slogan of the '60s tells of one of John Deere's major strengths, its network of independent dealerships. The blacksmith who founded the company recognized that a fair profit for the dealer was just as important as a fair profit for the company. Deere has consistently channeled its sales to farmers through these independent dealers, rather than making direct sales or using company-owned stores.

It would be difficult to have good dealers without good products to sell. It would be equally difficult to have good products if there were not good dealers to sell them. John Deere appears to have mastered this "chicken and egg" situation well. Most owners of the 1,800 John Deere dealerships in the U.S. and Canada are well-respected community leaders. Many are second generation, several are third generation and a few are even fourth generation John Deere dealerships.

After-sales service is vitally important to the customer. Dealer service technicians are well trained and properly equipped. Deere operator's manuals and parts catalogs are the envy of the industry. Operator's manuals are furnished with the equipment, and parts catalogs may be purchased by the customer. Fundamentals of service (FOS) and fundamentals of machine operation (FMO) manuals were developed for John Deere dealers and customers, but are used also as textbooks in high schools and some colleges worldwide.

One of the reasons that John Deere equipment brings higher prices at trade-in time or at farm sales is the ready availability of repair parts. Dealers stock the fast-moving parts and can get other parts quickly— generally overnight—from a parts depot. Many parts are still available for the popular Model "A" and "B" tractors made 50 years ago.

Here's a 2755 tractor with a Roll-Gard frame and canopy and 4-bottom 1000 integral plow. A center-pivot irrigation system, shown in the background, helps make this sandy river-bottom land more productive.

The 2755 high-clearance tractor uses front tires the same size as the rear tires used on tractors in the '50s. Mechanical front-wheel drive provides the traction needed to pull this special semi-trailer from soft fields. The trailer hauls the perishable crop directly to market.

This industrial yellow 2755 general purpose MFWD tractor differs from its farm tractor counterparts by having smaller diameter wheels and shell rear fenders.

A 2755 tractor with MFWD and ROPS is shown at the Geneseo, Illinois, dealership.

The 80-hp 2855N is the most powerful orchard/vineyard tractor on the market. It and the 2355N are narrow and less than 59 inches high (to the top of the steering wheel). Their front lights are low and narrow to minimize tree contact. A telescoping Roll-Gard ROPS is standard on both models. It can be lowered when working around low-hanging branches.

This 2955 tractor makes a minimum number of passes through the field in this conservation tillage operation using a 7200 MaxEmerge 2 planter with liquid fertilizer attachment.

In the setting sun, this 2955 MFWD tractor pulls a 215 disk.

The 96" front axle on this 2955 wide-track tractor provides 25" crop clearance for wide-row crops like tobacco and some vegetables. The tractor is shown with an S-tine field cultivator adjacent to a field of ripening flue-cured tobacco.

A 3155 tractor dwarfs a beautifully restored 110 lawn and garden tractor at Klug Farm Equipment, Eau Claire, Michigan. The 110 has the original 7-hp Kohler engine and is serial number 2991.

This 6-cylinder 85-hp 2955 high-clearance tractor has 4-post Roll-Gard protection.

This 95-hp 3155 tractor needs its standard MFWD at times when pulling this 8-row 7200 Ridge-Till planter in corn stubble.

The 4555 Tractor Joins the Waterloo Line

At Palm Springs, California, in January 1989, the largest new product announcement in the history of John Deere introduced the 55 series tractors from the Waterloo Works. Six models were divided into three with 106" wheelbase and three with 118" wheelbase and wider hoods. All were equipped with a redesigned 7.6-L engine. They were the 105-hp 4055, 120-hp 4255 with a Hi-Crop version, 140-hp 4455, 155-hp 4555, 175-hp 4755 and 200-hp 4955. The 4555 was a new intermediate model and the 4955 topped the line as the first John Deere row-crop tractor at 200 PTO hp.

The new engines proved highly fuel efficient. The 4955 set a new record for fuel efficiency at maximum power in the Nebraska tractor tests. All six models featured new valves and ports, higher top piston rings, automatically controlled viscous fan drive to match cooling requirements, and new 7-hole injector nozzles. The new high-tech turbochargers gave greater air-handling capacity on all models and the top two also were aftercooled.

The economical 16-speed Quad-Range transmission with Perma-Clutch was standard on all but the 4955. The 4955 had the ultra-deluxe 15-speed Power Shift transmission, which was optional on the other models. Caster/Action MFWD now featured a switch for automatic engagement/disengagement. The Sound-Gard body had increased visibility.

The three larger tractors featured electrohydraulic hitch control. All six models had a remote switch to permit the operator to slowly raise and lower the hitch from outside the tractor. Hitch lift capacity as high as 10,000 lb. was available as an option on the 4955.

This 4055 tractor is operating a 1508 rotary cutter. The photo was given to dealers at the Palm Springs introductory meeting; it was taken before the number of headlights was changed from four, as shown here, to the current three.

A 4255 tractor is shown with 630 disk. The black tool box is standard equipment and can be lifted off to keep the tools near the operator during adjustments or repairs.

Here's a 4055 MFWD tractor with optional radar, shown under the battery box. This unit provides the operator with a readout of percent of wheel slip, true travel speed and acres per hour worked.

This photograph was used to encourage farmers to trade up to the new 105-hp 4055 with MFWD from their 90-hp 4040 with 2WD. The styling is very similar on these two tractors with Sound-Gard body, even though they were built a decade apart. The Henry Dreyfuss Associates insist that changes in appearance for change's sake should be minimal. Continuity of design can increase resale value of used equipment.

This 4255 tractor, with 714 mulch tiller, shows the lean of the front wheels during turns with the 13-degree Caster/Action mechanical front-wheel drive.

This 4255 Hi-Crop tractor with cane and rice tires was part of the Waterloo Tractor Works exhibit of current tractors at Expo II in July 1990. Tens of thousands of people came to see more than 400 restored 2-cylinder tractors near the Waterloo, Iowa, airport.

Here's a 4455 tractor with triple rollers preparing a field for flood-irrigated rice. The 4455 has been the best selling John Deere tractor in its series, a distinction also held by its 4430 through 4450 ancestors.

A 4455 tractor with MFWD and front fenders is shown with a 714 mulch tiller.

IntelliTrak instrumentation on the 55 series tractors uses state-of-the-art electronics to monitor various tractor systems to increase productivity of the operator/tractor/equipment team.

The new 155-hp 4555 illustrates the front handhold to aid refueling, standard on 2WD models for the first time in all the 55 series Waterloo tractors. Also note the triple step. Safety and convenience continue to be major design considerations at John Deere.

The 4555 with MFWD and rear duals is well suited for heavy tillage work. It is the lowest-cost model offering the new electrohydraulic hitch.

This 4755 has dual cane and rice rear tires. The tractor carries a heavy load, with its two front-mounted sprayer tanks and the 8-row integral 7300 planter.

A 4755 2WD tractor with single rear wheels is used to accurately shape beds in sandy soil for an irrigated vegetable crop.

ADVERTISING YOU CAN COUNT ON

Word-of-mouth advertising by satisfied users of John Deere's steel plow multiplied his original business. Comments by satisfied users are still convincing, but today "the word" is spread in many ways.

Unlike most of its competitors, John Deere has consistently chosen to create most of its own advertising. Deere & Company advertising department does the complete job with professional copywriters, graphic designers, photographers, and production personnel.

Its advertising makes conservative claims that appeal to the intelligence of the farmers by showing them how they can make more money with John Deere equipment. Many of the copywriters have spent most of their careers at Deere, and many have a farm background. Their approach to getting prospect's attention is usually innovative, resulting in Deere getting a large share of awards in competition with other agricultural advertisers.

One of their approaches has traditionally appealed to the entire farm family. The John Deere dealer invites families in for John Deere Day. A movie shows the latest in farming practices and equipment.

The Furrow *magazine provides 2 million farm families with interesting and useful information on agriculture, a bit of humor, and advertising of John Deere equipment. It began in 1895 and has been continually published longer than any other farm magazine. It is also the most widely distributed, with circulation in more than 40 countries in 11 languages. Tailored to local needs, there are 12 editions in North America and 17 overseas.*

The leader of the line, the 4955, needs mechanical front-wheel drive and dual rear wheels to fully utilize its power for plowing.

Mechanical front-wheel drive is needed on this 4755 tractor for ample traction with the 915 V-ripper. About half of the 4755 tractors are sold with MFWD and over three-fourths of the 4955 tractors have this option.

A 4755 tractor with MFWD is shown during drawbar testing at the University of Nebraska at Lincoln in May 1989. The white "car" contains the instruments, and the following vehicles provide the variable loading. These tests are recognized worldwide for their accuracy and objectivity.

All-New 60 Series
4-Wheel-Drive Tractors

The 60 series 4-wheel-drive tractors were introduced in Denver in the fall of 1988 and were featured again in the Palm Springs new product meeting in January 1989. These articulated tractors were completely new in concept, design and appearance. They had an all-new longer-wheelbase chassis with center frame oscillation. The three new transmissions were the 12-speed Synchro, 24-speed PowrSync with built-in Hi-Lo, and 12-speed Power Shift. The 235-hp 7.6-L engine powered the 200-PTO-hp 8560, the 300-hp 10.1-L engine powered the 256-PTO-hp 8760, and a 370-hp 14-L Cummins engine powered the 322-PTO-hp 8960. The three models were similar except for the choices of engines and transmissions.

The Sound-Gard body featured a new side-access door and a one-piece upper windshield for improved visibility. The offset position of the muffler and air intake continues to minimize obstruction of vision.

With snow-capped mountains in the background, an 8760 tractor is at home with a 777/610 air seeder in Western wheat country.

The 4-wheel-drive 200-hp 8560 provides an alternative to the 2-wheel-drive 200-hp 4955 for tillage and other operations on large farms.

An 8760 tractor pulls a 9501 PTO-driven Maximizer combine in neatly windrowed wheat. The optional PTO on the 60 series tractors can be installed by the factory or the dealer.

This triple-wheeled 8960 pulls two 752 no-till drills in Western wheat hills.

Here's a new 8960 on triple tires, on July 23, 1990, at the Waterloo, Iowa, airport. It was part of the John Deere Waterloo Tractor Works display for Expo II, the big show of John Deere 2-cylinder tractors.

Tractors, Overseas

Germany, the Largest Overseas Tractor Source

Heinrich Lanz, A.G. built its first multicylinder gasoline tractors at Mannheim, Germany, in 1911. The famous single-cylinder Bulldog model appeared in 1921, the world's first crude-oil-fueled tractor. Several of these 12-PS tractors have survived, with one at the Deere & Company Archives in East Moline, Illinois, and another at the Mannheim Works museum. (PS is the abbreviation of the German word for horsepower and is usually net engine power. PTO hp, used in North America, is more conservative at about 85% of net engine power.)

A 4-wheel-drive articulated Acker-Bulldog tractor, produced from 1923-26, is on display at Mannheim. In 1926 the HR2 22/28-PS or Gross-Bulldog was introduced, with a 4-speed transmission. The plowing version had speeds of 2-5 mph and the road model, on solid rubber tires, provided 3-9 mph. In 1928 a 28-PS crawler version was added to the line. Sales took off with the introduction of radiator cooling for the 30-PS HR5 and 38-PS HR6 models, with more than 11,000 being built from 1929-1935. From 1952 to 1954 17-PS to 36-PS semi-diesel models were introduced. They were started on a mixture of gasoline and diesel by a unique rocking action of the electric starter.

After the acquisition of Lanz by Deere & Company, the first two tractors introduced were the 4-cylinder 10-speed 28-PS 300 and the 36-PS 500 early in 1960. They were followed in 1962 by the 2-cylinder 6-speed 18-PS 100 and the 4-cylinder 10-speed 50-PS 700.

The line was restyled in 1963 and broadened with the addition of the 303 and the 505. The Lanz name was then dropped, outside of Germany. The engine side screens were removed but the tractors retained the same engines. Shell fenders replaced the full fenders. The 2-cylinder model was upgraded to the 25-PS 200. All models retained the Lanz transmission, but the 3-cylinder 32-PS 310 and 40-PS 510 and the 4-cylinder 50-PS 710 now used a new engine from the Dubuque (Iowa) Works, the one that was later used in 20 series tractors.

Gus Gutslaff, retiree and acting curator at the Mannheim Works museum, shows a restored 300 tractor in 1989.

The final color of the Bulldog tractors before the acquisition by Deere is shown on the 40-PS D4016 and the 11-PS D1106 in the Mannheim Works museum. The Bulldog line became green and yellow in 1957-58, and the 19 diesel models were gradually reduced in number.

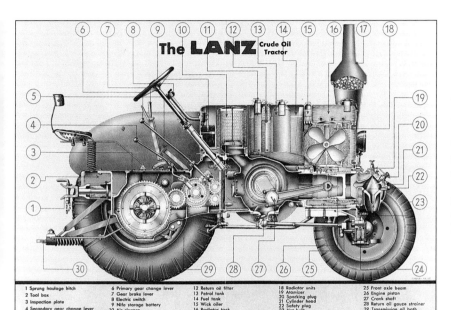

The LANZ Crude Oil Tractor

1 Sprung haulage hitch	6 Primary gear change lever	12 Return oil filter	18 Radiator units	25 Front axle beam
2 Tool box	7 Gear brake lever	13 Petrol tank	19 Atomizer	26 Engine piston
3 Inspection plate	8 Electric switch	14 Fuel tank	20 Sparking plug	27 Crank shaft
4 Secondary gear change lever	9 Nife storage battery	15 Wick oiler	21 Cylinder head	28 Return oil gauze strainer
5 Clutch pedal	10 Air cleaner	16 Radiator tank	22 Safety plug	29 Transmission oil bath
	11 Oil container	17 Silencer	23 Hot bulb	30 Sprung implement hitch
			24 Front axle spring	

The inner workings of the 45-PS single-cylinder D-9506 HR8 tractor are shown. Built from 1934-1955, it had a 6-speed transmission and a front axle with springs. Its 2-stroke engine ran on crude oil vaporized by spraying it on a hot bulb. The bulb was warmed up by starting on gasoline (petrol) similar to early John Deere tractors that ran on kerosene or fuel oil.

This 86-PS 3120 tractor, built in 1969 at Mannheim, was the first John Deere 6-cylinder tractor that wasn't built in the Waterloo (Iowa) Tractor Works.

① A 2020, built at the John Deere Mannheim Tractor Works, is working up a fine seedbed with a rotary tiller.

② This experimental 6-cylinder tractor was developed for Europe during the 10-series period.

197

Worldwide Tractors Introduced in Europe

The fall of 1967 saw the introduction of the 20 series in Europe. Five new sizes were announced: the 3-cylinder 34-PS 820, 40-PS 920, 47-PS 1020 with orchard and vineyard versions, and 52-PS 1120, and the 4-cylinder 64-PS 2020, also available as an orchard model. More sizes of utility tractors were required than in the U.S., due to the small farm sizes in Europe. The 3020, 4020 and 5020 were imported from Waterloo for those markets requiring larger tractors. Using similar ratings, the 4020 had 113 PS and the 5020 158 PS.

In 1968 the 72-PS 2120 was added to the line, and the following year saw Mannheim's first 6-cylinder tractor, the 86-PS 3120. Engines for tractors built in Mannheim were then and are now built at the John Deere factory in Saran, France.

The original 30 series tractor line, announced in Europe in 1972, retained the 20 series styling for the next three years. During that time the new OPU (operator's protection unit) was introduced on the 3130, and later on tractors down to the 1030 size. It was a factory installed module, built to meet all the anticipated European sound requirements.

The three largest models, the 2030, 2130 and 3130, were announced first, followed in 1973 by the 1630. The four smaller models were not introduced until 1975, when all eight basic models took the Waterloo tractor styling. As a result the specialty tractors remained the 920, 1020 and 1630 vineyard, and 1020, 1630 and 2030 orchard models.

Another specialty tractor introduced with the new styling was the 2030M or Multi-Crop, the high-clearance tractor for Europe. Hydrostatic front-wheel drive, already available on the Waterloo tractors, was offered on the 2130 and 3130. By 1977 the new-look models from the 1030 up had a better option added: mechanical front-wheel drive. This was a non-Deere design with offset drive, but was in increasing demand for European conditions. The last addition to the 30 series filled the power gap between the 2130 and 3130 in 1978 with the new 6-cylinder 3030.

① This 3-cylinder 51-PS 1130 has a purchased cab. Note the European headlights styled into the fuel tank and the front hitch pin in the base front weight. Front hitches are very useful for maneuvering trailers and implements into sheds.

② Here is the standard Mannheim 2030 tractor, a very popular mid-size tractor in Europe.

③ A pleasant European scene showing a 1040 with OPU and MFWD plowing near a farmyard.

A 940 tractor, with 4-post ROPS and canopy, seen here plowing.

A used 3130 with hydrostatic front-wheel drive and operator protection unit (OPU) is shown in front of a 3050 with MFWD and Sound-Gard body at a dealership in Gloucestershire, England, in 1990.

The European 30 series tractors offered a variety of options. The 830 and 930 have 2-wheel drive and 4-post, fender-mounted ROPS. The 1030, 1130, 1630 and 2030 feature mechanical front-wheel drive and factory installed OPU. The 2130 and 3130 have hydrostatic front-wheel drive and the OPU.

In 1975 the Mannheim-built tractors for the U.S. market had become the 40 series. However, it was not until the fall of 1979 that the "Schedule Masters" were introduced in Europe for the 1980 season. The whole line from 840 to 3140, including the specialty models and those built in Spain, changed on this occasion. In addition to their narrow width, the vineyard models could be distinguished from the orchard models by their headlights inset in the front grille. The orchard models retained their previous light position.

A significant development for Europe in 1981 was the SG2, an updated Sound-Gard body for the 1640 and larger European-made tractors; it was built in the new Bruchsal Works near Mannheim. The following year another cab, the low-profile MC1, was made available for stock farmers and others with low buildings. It was a safety cab of similar construction to the OPU, but with the lower roof line. It was an option on the smaller models.

To compete with other European manufacturers in a poor economic climate, the X-E and L-P series tractors were introduced in 1983. The four X-E models, from the 62-PS 1640 through the 82-PS 2140, were competitively priced, well-equipped tractors with the SG2 cab and MFWD options. The L-P series had the low-profile MC1 cab on the five 1040 through 2140 models used by livestock farmers.

The 6-cylinder 112-PS 3640 was introduced in 1984, complete with the exciting new multi-implement front 3-point hitch and PTO. The high-capacity front hitch permitted the use of two or more implements in combined operations. Like its U.S. cousin, the 3150, its standard equipment included MFWD as well as the 16-speed Power Synchron Hi-Lo transmission and 540/1000 rpm PTO with hydraulic control.

The low-profile Mannheim 1140 tractor took the place of the 1630 low-profile tractor in Europe and Canada. It was similar to the 2240 orchard/vineyard model in the U.S.

The 6-cylinder 3040 tractor has a noticeably longer wheelbase than the 3- and 4-cylinder models. Note the claw on the draft links, a European alternative to the Quik-Coupler hitch that is common in North America.

The new low-profile MC1 cab on an MFWD 1640 provided the livestock farmer with protection from the weather. It still permitted entry into low sheds when loading manure.

A 2140 with OPU cautiously descends a hill with a loaded 714A forage box on a tandem-wheeled trailer.

Here's a 1350 tractor and 175 loader with a good load from a stack of manure. Note the fender-mounted 4-post ROPS without canopy.

This 1550 tractor features the low-profile MC1 cab and illustrates the lean of the 12-degree Caster/Action mechanical front-wheel drive.

Tractors, Overseas

It was four years after the 1982 introduction of the Mannheim-built 50 series utility models in North America that the 50 series tractors were announced in Europe. The new models were the 3-cylinder 1350, 1550, 1750 and 1850; 4-cylinder 2250, 2450, and turbocharged 2650 and 2850; and the 6-cylinder 3050, 3350 and turbocharged 3650. They were still current in 1990, the end of the period covered in this book.

The last development introduced in 1990 was the CC2 Console-Comfort cab, a replacement for the MC1, offering a clear deck and entrance doors on both sides. A cushion-grip tilt steering wheel and console-mounted gearshift levers were among its innovations for an economy-priced cab.

This 1850N with MFWD makes effective use of its low height and narrow width in one of Europe's many vineyards.

TRACTORS ON FARMS IN 1987 IN SELECTED COUNTRIES	
World	25,597,504
United States	4,676,000
U.S.S.R.	2,736,000
Japan	1,904,070
France	1,519,760
West Germany	1,469,956
Italy	1,315,427
China	891,952
Brazil	780,000
Canada	742,200
India	697,968
Spain	678,680
U.K.	519,495
Australia	332,000
Argentina	208,000
South Africa	183,000
Mexico	163,000

Source: FAO Production Yearbook, 1988

Here's a 3050 hauling grain with a tandem-wheeled trailer. Most European tractors have a heavy-duty trailer hitch above and forward of the drawbar hitch. Trailers are used more frequently than farm trucks.

The latest addition to the 50 series lineup, a 1950 with MFWD and MC1 cab, wends its way homeward. Many European farmers live in villages and drive daily to their remote farm fields.

The large front tires on this turbocharged 4-cylinder 2850 are often needed for tillage and other farm operations. The more humid climate of Europe encourages farmers to work their fields wetter than in North America; compaction appears to be less serious.

A 3350 works with a 4-bottom integral 2-way plow fitted with slatted bottoms.

The front hitch and PTO on this 3650 tractor permit European farmers to do a variety of combined field operations. They can mount a plow on each end of the tractor, combine secondary tillage with planting, or perform two haying operations at once.

Spanish Tractors for Local and Special Needs

Lanz began building a variety of sizes of diesel tractors from 11 to 40 PS in Germany in 1953. When Lanz opened the new Getafe Works near Madrid, Spain, in 1956, the 40-PS model was the first tractor produced. The 505 was the first Deere-designed tractor to be built in Spain in 1963, followed by the 515, 717, 818, and an orchard/vineyard version of the 515 some three years later.

Spain announced their 30 series tractors in 1973, following Mannheim's similar models announced in 1972 and 1973. In 1977, Spain introduced the 35 series for their own market and for export to other Spanish-speaking countries. The switch to the 40 series came in 1980 with both the regular models and some specialty models having low sales volume. Tractors produced by John Deere in Spain have generally been based on the design of models built in Mannheim, but with modifications to more nearly match their market or export markets for specialty tractors.

The first tractor produced by Lanz in their Spanish factory in 1956 was this 40-PS Bulldog. The company was named Lanz Iberica-S.A., as Spain is the largest country in the Iberian peninsula. Note the clean lines of this single-cylinder tractor, and the 3-point hitch. The crankshaft location is similar to that of John Deere 2-cylinder tractors.

The 818 was unique to the Spanish market because it didn't have a Mannheim equivalent. Other Spanish-built tractors were the 515 and 717.

The 6-cylinder 3135 uses both its traction and hitch lift capacity to operate this 4-bottom 2-way Spanish-built plow.

The Spanish 1040V, illustrating its built-in headlights and special fenders, is working in a vineyard. Other specialty models from Spain were the 1140V and the 1140F, 1640F and 2040F orchard, or fruiteros, models.

Here's a 2450F orchard, or fruiteros, model at the Getafe Works in Spain in 1989.

The 1140M high-clearance or Multi-Crop model provided extra clearance for certain European crops. It was intermediate in height between the U.S. Hi-Crop and HU or high utility models.

Large Tractors Sourced From the U.S.

One of the early imports to Europe from the U.S. after the 2-cylinder days was the 1010 agricultural crawler. A significant number of 3020, 4020 and 5020 tractors also were imported from Waterloo for larger European farms.

In 1981, John Deere introduced a very important feature to European farmers on two tractors imported from Waterloo and assembled in Mannheim. Caster/Action mechanical front-wheel drive was introduced in Europe a year before it was announced in the U.S. The European 4040S and 4240S had turbocharged 6-cylinder engines, while the U.S. 4040 and 4240 models had naturally aspirated engines.

Waterloo's 4250 through 4850 tractors were introduced in Europe in 1984. A special tractor for Europe was introduced in 1985—the turbocharged 6-cylinder 140-PS 4350 with Caster/Action mechanical front-wheel drive as standard.

INDUSTRY TRACTOR SALES IN 1977 IN SELECTED COUNTRIES

United States*	154,923
Italy*	72,276
West Germany*	64,184
France*	62,209
Brazil	59,249
India	33,146
U.K.*	33,000
Spain	30,774
Canada	27,982
Argentina	23,739
Australia	21,588
Mexico	16,767
South Africa	15,000

Source: *Agricultural Machinery Journal* May 1979
* Also a major exporter

This 4020 diesel was rescued from the weeds and completely restored by Roger Perry of Ticknell in Derbyshire, England. The tractor was shipped to the U.K. in December 1971. Its serial No. 263,524 places it in the last 4 percent built of this classic tractor.

The operator's view of Perry's 4020 Power Shift tractor shows the pleasing and practical design developed by the Henry Dreyfuss Associates.

Some of the differences between tractors sold in North America and Perry's 4020 tractor sold in the U.K. are visible here. The front axle is in the longer-wheelbase row-crop position. The rear fenders are similar to those on U.S. standard tractors but they do not join the platform. Headlights are in the position of U.S. models without rear fenders.

Here's a 4040S tractor fully weighted, front and rear, and with lighting to meet European road requirements. The front hitch pin is available with or without the suitcase weights.

A 4240 with hydrostatic front-wheel drive makes easy work of plowing with a 6-bottom semi-integral Ransomes plow near Gloucester, England. North European plows are generally lighter weight than U.S. plows of similar width, because soil conditions are more favorable.

This European 4240S follows the return furrow with a 2-way plow. Fourteen front suitcase weights balance the plow in transport and make the MFWD more effective.

Small Tractors Sourced from Italy

In 1987 John Deere extended its specialty orchard/vineyard and small tractor line in Europe by obtaining a few models of tractors that Goldoni was selling. Engines built in Saran, France, were used and some modifications were made for John Deere styling.

Innovations for orchard work include retractable lights, a telescoping steering wheel, and an optional cab with 4-pin easy removal. An 8-speed transmission has 19-mph road speed. Options include a 16-speed creeper transmission and mechanical front-wheel drive.

Tractors in South Africa

After having a one-half interest in a local South African company since 1930, John Deere began manufacturing in 1962 in its Nigel Works in the Transvaal. Tillage machines were made and some tractor models from Mannheim were assembled. The larger Waterloo models were imported assembled.

Starting in 1982, the South African government insisted that Perkins engines, built by the new Atlantis Diesel Engine Company, be used in all locally assembled tractors. The resulting 41 series tractors were similar to the European 40 series and proved themselves by obtaining 17% market share.

In September 1987 they were replaced by the current 51 series, with six sizes and 16 models. These were the 4-cylinder naturally aspirated 2251 and 2251N, the altitude-compensated 2351, 2651 and 2651TT TransTill model, the turbocharged 2951, and the 6-cylinder compensated 3351 and 3651. The altitude compensator packs more air into the combustion chambers when working in the Highveld. All 51 series tractors have the TSS Mannheim transmission, with eight forward and four reverse speeds on the two smaller models, and 16 forward and eight reverse speeds on the four larger. All models offer the 12-degree Caster/Action MFWD option. The two larger models have a 2-post Roll-Gard frame as standard equipment; on the four smaller models it's an option.

This 34-PS 938RS tractor is one of seven compact 4-wheel-drive models, with equal-size wheels, built by Goldoni in Italy. The other three rigid tractors are the 29-PS 933RS, 34-PS 1038 and 38-PS 1042. The three articulated models are the 18-PS 921, 29-PS 933 and 34-PS U238.

The 2951 is the largest model assembled in South Africa with a 4-cylinder Perkins engine.

A 48-PS 1745F tractor with rear-mounted forklift loads a trailer pulled by a turbocharged 3-cylinder 67-PS 2345F. Other Goldoni-built orchard models are the 42-PS 1445F and 56-PS 1845F. Note that the standard folding ROPS is raised on one tractor and lowered on the other.

The 3351 has a 2-post Roll-Gard ROPS as standard equipment.

Mechanical front-wheel drive on the 2251N orchard tractor assists in pulling a heavily loaded sprayer in hilly terrain.

Argentine Tractors Produced in Rosario

From the opening of the John Deere Rosario Works in Argentina in 1958, the 2-cylinder 730 diesel model was produced. Four styles were offered: standard, tricycle, adjustable front axle and Hi-Crop. More than 12,000 of these were built by 1965 and they were produced until 1970.

In 1963, the horizontal 2-cylinder models were joined by the vertical 2-cylinder 2-stroke diesel 445. This was the Argentine version of the U.S. 435, which used a General Motors engine. It also was available in four styles; RU regular, T tricycle row-crop, O orchard and V narrow for vineyard work.

In 1970 the 2-cylinder models finally gave way to the 20 series with U.S. styling. The 1420 had a 3-cylinder 43-hp John Deere engine with the 445's 5-speed transmission. The engine of the 4-cylinder 66-hp 2420 was similar to the Mannheim 2120, and the 6-cylinder 77-hp 3420 was similar to the European 3120. The 4420 used a 6-cylinder 102-hp engine from Saran, France. All three larger tractors were fitted with 8-speed Syncro-Range transmissions to simplify tooling and provide better availability of repair parts. The 20 series were built from 1970 to 1975, when they were replaced with the 30 series—2330, 2530, 2730, 3330, 3530 and 4530.

These in turn gave way to the 40 series in 1981, reduced to the 2140, 3140 and 4040 models, due to the financial climate in Argentina. In 1984 the 4040 became the 4050 for two years, but was then replaced with the 3540 for another two years until the whole line was updated. The 50 series consists of the 4-cylinder 95-engine-hp 2-wheel-drive 2850, the 6-cylinder 110-hp 2WD or MFWD 3350, and the 6-cylinder turbocharged 125-hp MFWD 3550. All three current models (as of 1990) have engines with similar bore and stroke. They use the 8-forward, 4-reverse-speed transmissions introduced on the 40 series tractors.

The 2-cylinder 2-cycle General Motors diesel engine is shown in the 445 tractor with rear-mounted cultivator.

The Argentina-style medallion is shown on the front of a 445 vineyard tractor. JD was used in place of the leaping deer to avoid a trademark problem.

The 1420 had a current John Deere 3-cylinder engine but retained the 445 transmission. Countries with limited sales can't justify the variety of powertrain options available from Europe or the United States.

The 4-cylinder 2420, shown cultivating soybeans, was similar in appearance to the 2520 in North America.

Here's a 3540 with mechanical front-wheel drive, shown with optional industrial yellow paint.

This 6-cylinder 3350 is taking a broad sweep with a locally built 7-bottom plow.

John Deere Leads in Mexico

In 1957 Deere & Company started a sales branch and bought a site for a factory in Mexico. After importing the 2-cylinder 620 and 720 tractors from the U.S., Deere began assembly of the 435, 630, 730 and 830 tractors in the new factory. Then came the 3010 and 4010, followed in the 1964 season by the 3020 and 4020. In 1966 the line was joined by the 1020 and 2020, and in 1971 by the 2120 from Europe.

The new 35 series tractors were introduced in 1973, with the 4435 followed by the 4235. These were followed in 1975 by the 2535 and 2735, with all four staying in the line until 1983, when they were replaced by the Mexican-built 55 series.

John Deere celebrated its 35th anniversary in Mexico in 1990. It is the largest manufacturer of tractors and implements in the country, with factories in Monterrey and Saltillo.

The 4235 tractor and integral 2-way disk plow built in Monterrey were popular among wheat farmers in the north and western areas of Mexico.

This 2755 Turbo tractor illustrates the good crop clearance under the front differential. All Caster/Action mechanical front-wheel drives have the final reduction in the front wheel hubs to permit using smaller differentials and universal joints.

The 2535 tractor has adequate clearance for cultivating corn, an important crop in much of Mexico.

The local manufacturer used a flared exhaust pipe above the muffler on this 4435 tractor. The offset disk was made by John Deere in Monterrey, Mexico.

A 2555 tractor transports an MX221 integral disk.

This 2555 low-profile tractor is working in a Mexican orchard.

John Deere Joins Chamberlain in Australia

From an idea for a tractor design conceived in the 1930s and two prototypes built in 1948, Chamberlain became the largest producer of tractors in Australia. Chamberlain's first diesel, the 60DA, was introduced in 1952, fitted with a 3-cylinder 2-cycle GM 3-71 engine. These more expensive but well-accepted tractors stayed in production until 1967. A smaller model was required in addition to the Champion and the Countryman shown. About 525 Canelander tractors were built in 1959-66 with 48 PTO hp, powered by imported Perkins engines.

The merger with Deere took place in 1970, and in 1975 the 80 series Chamberlain tractors were announced, with many John Deere features. John Deere engines from Saran, France, powered the 4-cylinder 68-hp 3380 and the 6-cylinder 85-hp 4080, 98-hp 4280, and turbocharged 119-hp 4480, all governed at 2200 rpm. With their side radiator air intakes and their 6-post ROPS or safety cabs, they were acquiring a family likeness. In 1983 they were redesigned somewhat to become the 80B series, still with the Chamberlain name and finish.

The green and yellow 90 series Chamberlain tractors appeared in 1982. The 94-hp 4090, 110-hp 4290, 129-hp 4490 and 154-hp 4690 all used John Deere 6-cylinder engines; the three larger tractors were turbocharged. All had 12-speed collar-shift transmissions with Hi-Lo. Caster/Action mechanical front-wheel drive was an option on all four. The operator's compartment was even more like the Sound-Gard body, with a slightly curved windshield and similar roof. External lift cylinders were standard on the three larger models, optional on the 4090. All could have up to four remote outlets.

In May 1986 tractor production in Welshpool was terminated in favor of importing complete tractors from John Deere. The factory in Perth continues to build tillage and seeding equipment for the regional market. Deere has become the farm equipment market leader in Australia and New Zealand, with 200 John Deere dealers in the two countries.

Here's a Chamberlain 40K tractor at a demonstration pulling a disk plow in 1949. More than 2200 of these horizontally opposed 2-cylinder kerosene tractors (40K, 40KA, 45KA and 55KA) were built over the next eight years.

The 4480B tractor had a distinctly industrial look, with a square nose and Chamberlain's distinctive yellow color.

The 1955 Champion 6G by Chamberlain had a synchromesh transmission, a wide seat and a front bumper. A Perkins diesel engine powered the 57-PTO-hp tractor. More than 20,000 were built in the next 20 years.

The needs of large farms were met, starting in 1958, with the Countryman model, shown with dual rear wheels. A Meadows engine imported from England provided 75 PTO hp. The sprung front axle was also popular in Europe and is similar in principle to that used on Ford cars in the United States in the '30s.

The smallest of the 90 series, the 4090, is chisel plowing. It had a 6-cylinder 300 series engine from Saran, France.

The Chamberlain 4690 tractor in green and yellow looked quite similar to the John Deere tractors that later replaced it. This tractor was well equipped for the difficult tillage conditions of Australia with its turbocharged 400 series engine from Waterloo, Iowa, mechanical front-wheel drive, and dual rear wheels.

Implements

Matched Equipment

John Deere's success in the market is partly due to offering farmers a full line of tractors and equipment matched to their needs.

Matching of equipment to tractors has been one of the company's greatest challenges and strengths. The challenge comes from its decentralized organization, a long tradition. Each factory has its own product engineering department. Thus, the company depends on cooperation and coordination to have the right equipment available when new tractors are announced. The freedom to make decisions locally and at lower levels has recently been recognized as a need for the entire U.S. manufacturing industry.

Matching tractors and equipment to economic conditions on the farm is a unique challenge. Needs can change drastically between the time the equipment is planned and when it is produced. The '70s provided several years of good farm income, based on high grain export sales. During this time many farmers bought their first tractor with a cab. They also chose options like Power Shift, differential lock, dual wheels, Quik Coupler, and more remote hydraulic outlets. This ability to buy was greatly curtailed in the '80s and may not have been back to normal in 1990. One of John Deere's solutions has been to offer various packaged tractors, especially in the utility models. The base package has few options. Other packages include the most popular options for various regions of the market.

The hard times of the '80s reminded many people of the '30s and early '40s. In the mid '30s most John Deere tractors sold were more than 20 hp, with the Model "A" as the best seller. With a stagnant economy and more farmers wishing to switch from horses, the Model "B" became the best seller in 1938. John Deere introduced the 13-hp Model "H" for 1939. In 1939-41, models of 20 hp or less accounted for more than two-thirds of John Deere tractor sales.

Matching equipment to the size and type of prime mover is essential for dependable performance. Equipment for first-time tractor buyers had evolved from horse-drawn designs, and was usually no more than double the former size. However, in 1960 the New Generation of Power arrived with four models spanning a range of 35 to 84 hp. Each next larger model had 30-40% more power. In 1990 John Deere provided 15 sizes of farm tractors ranging from 45 to 322 hp. Succeeding sizes in the utility and row-crop models averaged 15% more power. Power increases between adjacent 4-wheel-drive models were less than 30%.

Self-propelled combines introduced another form of prime mover in the '40s. The advantage of not knocking down any standing crop when opening a field led to self-propelled cotton pickers, windrowers, and forage harvesters. Hi-Cycle sprayers provide clearance for standing crops that is not available from tractors.

Equipment must be matched to a variety of soil conditions and moisture. Mobility of combines in soft, wet rice fields used to be met with tracks. These have been replaced by cane and rice tires, and hydrostatic rear-wheel drive. Wider tires or dual tires provide better flotation for tractors. Mechanical front-wheel drive assist provides up to 20% additional traction. Dry soils require their own unique solutions, too. Hoe-type press-wheel drills work in heavy residues with minimum disturbance of the scarce moisture needed for germination.

Higher yields from better varieties and use of more fertilizer have increased crop residues. These were buried in the '60s with moldboard plows. Agricultural research and farmer experience have shown that it pays to leave more crop residues on the surface, especially during the winter. This shift in cultural practices has reduced the use of moldboard plows, which invert the soil. There is increased use of tine-type tillage tools such as rippers, chisel plows, field cultivators and combination tillage tools. Demand has increased for planting and cultivating equipment that tolerates more trash.

Tillage equipment is rather universal for most crops. Some crops should be planted by row-crop planters and others by grain drills or some variation of the two. Soybeans are planted by planters if they are to be cultivated, and normally by drills if they are not.

Harvesting is even more specialized by crop. Although the combine is rather universal, it has different heads available for windrowed wheat, standing wheat, rice, soybeans and corn. Cotton is even more demanding in its harvest. If it is a variety suitable for single-pass harvesting, a cotton stripper is used. If it is a higher yielding multiple-pass variety, a cotton picker is used.

Hay requires an even greater number of options for harvesting, as it is cured in the field and thus requires multiple passes. In 1960, most hay was cut with a 7' sicklebar mower. In 1990, mower/conditioners were very popular. Self-propelled windrowers offer another option. In the '60s most hay was put up loose or with square balers. In the '80s the large round baler became the best seller for harvesting hay because it reduced labor. Twine-tie square bales are still popular with dairy farmers, and wire-tie square bales remain popular for shipping hay. Forage harvesters have heads for standing hay, windrowed hay, and for corn.

Row spacings have narrowed, with 38" being the most popular in the '60s and 30" the most popular in the '80s. Soybeans show a distinct yield advantage in narrower rows, and corn shows some. Switching corn row widths involves three items of equipment, as the cultivator and corn head must match the planter with the same number of rows, or one-half or one-third the number. Many farmers change all three at once, a considerable

investment, though many cultivators and some planters can be adjusted to narrow or wide rows. Since the trend toward narrower rows for soybeans continues, planters are now available down to 15" rows. Grain drills have Tru-Vee openers available for better seed placement of soybeans. Cotton joined the narrow-row trend with the introduction of the 9960 cotton picker for either wide or narrow rows.

U.S. FARM EQUIPMENT

	Equipment on Farms 1987 Census	Industry Retail Sales 1989
Wheel tractors	4,603,870	106,570
Moldboard plows	2,600,000 *	2,811 **
Planters	1,160,000 *	68,343 ***
Grain drills	1,013,000 *	11,405 **
Cutterbar mowers	1,400,000 *	5,047 **
Mower/conditioners	651,970	13,152
Side-delivery rakes	1,000,000 *	10,925 **
Hay balers, <200 lb.	822,657	7,123
Forage harvesters	271,000 *	2,801
S.P. combines	667,056	9,110
Corn heads	368,000 *	
Manure spreaders	935,000 *	10,664 **

Sources:
Census Bureau, *1987 Census of Agriculture*
Equipment Manufacturers Institute

* *Implement & Tractor*, October 1987 (Estimates)

** Census Bureau, *Current Industrial Reports*, MA35A(89)-1 (Shipments)

*** CIR MA35A(89)-1 (Shipments of row units)

Farm tractor sales in the U.S. were less than 2.5% of the tractors on farms. Replacement rates for 11 important implements were even lower.

Tillage

Plows

The plow is mentioned in the Old Testament and is pictured on ancient Egyptian monuments. Originally a forked stick pulled by men, it is the tool that made the cultivation of crops practical. Later oxen were domesticated to pull plows.

Centuries later, in the 1600s, British landlords tried to improve on the design of the wooden plow. By the early 1700s a plow with a curved moldboard was introduced in England. This marked the beginning of plows designed to turn or invert the soil. This was a distinct improvement for northern Europe, with its humid climate and rapid growth of vegetation. At that time many plows were made entirely of wood, but some had a steel chisel point. Cast iron plows were made in Scotland as early as 1763. John Deere got his start in 1837 with the making of a steel plow, which scoured better in prairie soils.

Tractor plows of the '30s evolved from horse-drawn gang plows. Drawn plows accounted for most of John Deere plow sales until the mid '50s, when draft-sensing 3-point hitches were added to Deere tractors. Drawn plows used safety-trip hitches to protect the plow when stumps or rocks were hit. Tripping an entire integral plow was impractical, so John Deere engineers developed the first safety-trip standards for these plows in 1950.

By 1960, drawn plows had declined to only one-third of plow sales, with the 3- or 4-bottom 555 the most popular drawn plow. The 555 and 666 drawn plows were replaced by the F620-F620A in the '60s and the 3600 in the late '70s. The popularity of drawn plows continued its slow decline and in 1990 only accounted for one-sixth of John Deere plow sales.

Here's a 555 plow drawn by a 730 diesel in Kansas. It features a bolted truss frame and optional safety-trip standards. The 555 was a very popular plow in the '50s, and its sales in 1960 were three times those of any of its successor drawn plows.

This 3-bottom F45 integral plow was the popular choice for utility tractors for more than 15 years. It featured the NU (new universal) bottoms, designed for higher plowing speeds. The F45 had safety-trip standards, and the more economical F35 used shear-bolt standards.

The 3600 drawn plow was available in 5- to 8–bottom sizes, and could be hitched in tandem for up to 16 bottoms with 4-wheel-drive tractors. Safety-trip or spring-reset standards could be shimmed to provide 16-, 18-, or 20-inch width of cut.

This F125 4-bottom integral plow featured safety-trip standards and a welded box-beam truss. Its strength was matched to row-crop tractors. It accounted for over half of Deere integral plow sales from 1964 through 1969.

The 1000 integral plow, sold in the '80s, is one of the cleanest plow designs John Deere has ever made. The safety trip is clever, simple, and retains its trip resistance throughout the life of the plow. The plow is simply raised to reset the trip.

Integral (3-point hitch) plows continued their dominance only through 1964. In 1960 the best seller was the 810A, used with Waterloo 2-cylinder tractors. A close second was the 415A-416A, used with Dubuque tractors. The F120, used with row-crop tractors, took over as the market leader of integral plows. The F35-F45 was the leading integral plow for utility tractors until it was replaced by the 1000. Integral plows had two-thirds of the market in 1960 but had declined to less than one-sixth of sales in 1980, and the decline continues.

The big news from John Deere in 1962 was the introduction of a third way to attach a plow to a tractor. The semi-integral F135-F145 plow quickly captured the market. Sales of this plow in the U.S. and Canada exceeded 15,000 units in 1966. That year more than 25,000 John Deere plows were sold, the record for the 1960-90 period. Its successor, the F1350-F1450, was the best seller in the '70s until the 2600 and 2800 took over. Semi-integral plows continue to account for most sales of one-way moldboard plows.

Two-way moldboard plows are popular in the irrigated Southwest because they leave no dead furrows or back furrows. Their sales have held up better than those of other plows, accounting for little more than 5% of all plows in the '60s but growing to more than 20% in the late '80s. Most of these integral plows rotate sideways for the return trip. The 825 was the most popular unit in the '60s, followed by the 4200 in the '70s and '80s.

The 475 integral was the best selling disk plow in 1960, but less than 400 units were sold that year. The F215H 2-way disk plow took over as the sales leader, but all production of disk plows in the U.S. was dropped in the '70s. John Deere still makes disk plows in Monterrey for some of the hard soils of Mexico.

The 2810 semi-integral plow is the first choice for most Corn Belt farmers. Its hydraulically adjustable cut permits leaving up to 25% of the residue on top to control erosion. In-furrow plowing is available with four, five or six bottoms and on-land plowing with six, seven or eight bottoms.

Here's the spring-reset standard available on John Deere plows. It raises over large stones or stumps and automatically resets without stopping the plow.

This 2000 plow is the economical version of a semi-integral plow. It provides more than 29" of under-frame clearance. The four, five or six bottoms can be manually adjusted for a 14-, 16- or 18-inch width of cut. Like all semi-integrals, it adds weight to the hitch for improved traction.

Semi-integral plows trail like a shadow, as shown by this 1450 following a 4430 tractor. The rear furrow wheel is steered by the hitch crossbar. The plow featured a 6" square main truss tube and a rear gauge wheel.

Here's the most popular pair of the '60s, the F145H semi-integral plow and the 4020 diesel tractor. The semi-integral design results in a simple-to-hitch, medium-cost plow.

One-half of the plows on U.S. farms are in the adjoining states of Iowa, Minnesota, Illinois, Wisconsin, Ohio, Missouri, Indiana, Michigan and Kentucky. Their concentration parallels that of tractors, with about one plow per two tractors in most states.

In the '60s and '70s, farmers saw the trash-free surface of freshly plowed fields as potential money in their pockets. U.S. industry shipments peaked in 1966 with 110,000 plows. Shipments remained above 30,000 units a year during these two decades. The poor economy, combined with a recognition that some trash should be left on the surface, resulted in industry shipments dropping to less than 2,000 plows annually for 1986-88.

This F835A 2-way plow is mounted on a 4020 diesel tractor. Several front-end weights were required to balance the eight bottoms. The vane-type rollover cylinder had its origin in the power steering of 2-cylinder tractors.

The bottoms on the 3945 integral 2-way plow, introduced in 1989, resemble the first steel plow made by John Deere. In its second year of sales, it was John Deere's best selling model.

JOHN DEERE EQUIPMENT SALES U.S. AND CANADA IN 1975	
Farm tractors	50,522
Drawn plows	1,248
Integral plows	1,863
Semi-integral plows	7,485
Two-way plows	1,350
Total moldboard plows	11,946
Chisel plows	4,581
Bedders	713
V-rippers	993
Disks	14,159
Field cultivators	4,779
Roller harrows	1,372

This 4200 integral 2-way plow has been expanded to four bottoms. Standards could be mounted for 16- or 18-inch cut. A 3x6-inch-stroke cylinder provided positive rollover without latches. The 4200 was designed for tractors of 100-150 hp.

This 3945 plow has been reversed for the return pass. The reversing mechanism is similar to, but simpler than, the one on 2-way disk plows. This clever design from an outside inventor is much lighter than previous designs, reducing required hitch lift capacity and needing fewer front weights. The plow is made in Monterrey by John Deere for the U.S. and Mexican markets.

Here's an F210 reversible 3-bottom disk plow. At the end of the field, the main beam swings to the left, the disks are re-angled and the tail wheel is reversed.

Chisel Plows

The chisel plow has taken over much of the primary tillage previously done by moldboard plows. History appears to be repeating itself. The ancient wooden plow left a furrow like the chisel plow but at the shallower depth of a field cultivator. Many farmers in developing nations still depend on animal-drawn wooden plows. Although India makes and sells 100,000 tractors annually, most of their farmers were still using wooden plows in 1990. A wooden plow made in India in 1971 may be seen in the Deere & Company Administrative Center.

In 1960, most chisel plows were used in wheat rather than in corn. Twisted double-point chisels 3 inches wide were the most popular, but 2- and 4-inch double-point chisels were available. Sweeps were offered from 6 to 20 inches wide.

As chisel plows became more popular in the Corn Belt, they were designed to provide better trash clearance to avoid plugging. Vertical clearance was increased as well as the fore-and-aft clearance between the standard bars. Chisels are operated deeper for corn than for wheat.

Chisel plows offer several advantages over moldboard plows. Of immediate interest to the farmer is the ability to cover more ground with the same size tractor. This saves precious time in the fall and also saves fuel. For the longer term, keeping the trash on the surface reduces wind and water erosion. The grooves cut by the chisels also allow rain and snow to seep down into the subsoil.

The 100 series chisel plows were the best sellers in the '60s, with average annual sales of 2,000 units. The 1600 integral and 1600 rigid drawn chisel plows were the most popular units in the '70s. Total chisel plow sales exceeded 4,000 units annually in 1973-1976. The 1610 series took over in the early '80s, followed by the 610 chisel plow. Sales of chisel plows, which had been one-tenth those of moldboard plows in the mid '60s, caught up with moldboard plow sales in the last half of the '80s.

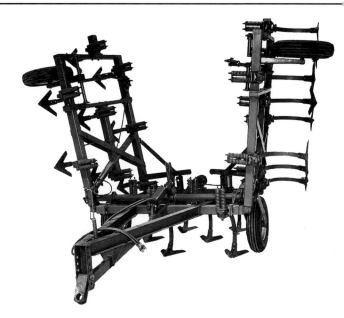

The wings on this 150F chisel plow have been raised for transport by the 10 hand winch. In operation, the wings are free to float with the terrain on gauge wheels. Twin springs in front of the bar permit the sweeps to rise over rocks.

Here's a 2-bar 1600 mounted chisel plow. It features a new, simpler and better trip design. The 1600 mounted chisel was the all-time most popular John Deere model. It was available in 8- to 20-foot sizes.

This 1600 rigid drawn chisel plow is leaving a rough surface with considerable residue on top. This helps keep blowing snow on the field. Drawn rigid 1600 chisel plows came in 8- to 19-foot sizes by use of frame extensions. The 3-section 1600 flexible folding chisel plow came in 20- to 41-foot sizes.

This 610 chisel plow features good trash clearance for cornstalks, clumps of bean trash or tangled cotton stalks. It has up to 32-inch under-frame clearance. Standards are spaced 36 inches apart on bars spaced 40 inches apart. The optional Tru-Depth spring-reset standards shown apply 1250 pounds of trip resistance and clear 11-inch rocks.

Other Primary Tillage Tools

A variety of other primary tillage tools have entered and departed the market over the years. Bedder-listers have been important tools in irrigated corn areas like Nebraska. In 1961 John Deere offered one front-mounted bedder, two hitch-mounted bedders and two drawn units. The integral and drawn units permitted the use of listers to plant and fertilize in furrows or on beds.

Disk tillers have been popular as a once-over tillage tool in the low-rainfall spring wheat areas of the prairies in Canada, the Dakotas and Montana. They require less power than either moldboard or chisel plows, and leave a soil surface with lots of trash on top. Their efficient use of time and fuel is further improved by adding the seeding attachment.

The V-ripper of 1990 is a far cry from its ancestors of the '60s. The 1961 issue of the *Modern Farming* catalog listed six models of subsoilers made by three John Deere factories. Each had a single standard, with maximum penetration of 16 to 26 inches. Some were obviously old designs, as the 20 drawn subsoiler had steel wheels and rope-trip lift. Their sales were generally limited to areas with recognized hardpans.

The 90 subsoil chisel plow and its successor, the 900 V-ripper, each had annual sales of more than 1,000 units in the '70s. Sales of V-rippers in 1990 approached those of all one-way moldboard plows. The V-ripper has gained popularity in the Corn Belt as a fall tillage tool to conserve soil and water. It also loosens compacted soil for better root penetration.

The mulch tiller (combination disk-chisel) has gained in popularity since its introduction in 1972. Its sales in 1990 almost matched those of chisel plows. The front-mounted disk reduces the length of stalks and eliminates most plugging of the chisel plow in the rear. It is a good primary tillage tool for fall or spring.

This double-bar F871 bedder is working in sod. It is planting in the furrow and applying fertilizer, insecticide and herbicide. Row crops required shredding before using the bedder.

The popular 1200 series Surflex tillers came in widths of 8 to 20 feet. Sections 4 feet wide gave the tiller flexibility over uneven terrain. The tiller worked the soil 2 to 6 inches deep with 18- to 22-inch disk blades on 8-inch spacings.

The 915 V-ripper uses a tubular frame to mount five to 13 standards in widths of 8 to 22 feet. Its 35-inch clearance allows it to rip 2 feet deep without plugging. Optional coulters help in heavy trash. Rippers should be used with 4455 through 8960 tractors.

Here's a safety trip protecting a standard from a hidden obstruction. In rock-free soils, the shear-bolt standard is adequate.

This 714 mulch tiller is the ideal tool to work directly behind the combine in heavy cornstalks. It can be equipped with a front coulter gang or the optional disk gang.

Disks

Disks have long been a favorite tillage tool of farmers. Their rolling action results in minimum draft for the width and depth of soil tilled. When oats were grown for horses, the oats were broadcast and then disked in as the only tillage operation. Heavy plowing disks, with wider spacings and more weight per blade, have been used for primary tillage for cotton, corn and soybeans.

Traditionally, disks were used in the fall to cut cornstalks ahead of the moldboard plow to prevent plugging and get better stalk coverage. In the spring, a field was disked once or twice before planting. More recently disks have been used to incorporate fertilizers and pesticides for more effective use of the chemicals.

Soil penetration and draft were adjusted on horse-drawn single disks by changing the angle of the disk gangs. Tractor-drawn tandem disks kept this same method through the '50s. As wheel lifts were added for transport, they also became gauge wheels during operation. Customers gradually learned that little or no adjustment of gang angle was needed.

In spite of the many uses for disks, industry shipments in the U.S. dropped from 93,600 disks in 1960 to only 13,234 in 1989. The two major influences on disk shipments and sales were the overall farm economy and the general reduction in tillage. Little of the change was due to profitability of any particular crop, because disks are used for most crops. There also has been some shift from disks to field cultivators. John Deere sales of field cultivators passed disk sales in 1984.

John Deere sales are also influenced by how well they match the equipment needs of the farmer relative to competition. Deere disk sales in the '60s paralleled plow sales and were more than 20,000 units in 1960-67. More than 9,000 RW-RWA wheel disks were sold in 1960. Other popular models in the '60s and early '70s include the AW-AWS-AWR, BW-BWS, BWF and KBA.

Here's a KBA disk in corn ground. The design was derived from an earlier drawn disk with a wheel lift added. It still permitted easy adjustment of the gang angles. The four gangs were free to flex independently.

The RWA disk was John Deere's first disk designed for wheel transport and gauging. The angle of the disk gangs was fixed. Weight pans were built in to add extra weight for penetration in hard soils. White iron bearings were standard and ball bearings optional. Antifriction bearings had limited durability in the dirty environment of disks until triple-lip seals were developed.

The BWA was a rugged disk with rectangular tubing gang frames. Limited adjustment of gang angle was provided. It was available in widths of 11 to 20 feet. The BWF was similar but included spring-cushioned gangs.

This 210 rigid disk was the favorite in the mid '70s. With 9-inch blade spacing, it offered the stalk-chopping ability needed by Corn Belt farmers. The 11- to 14-foot disks were matched to 4230 and similar tractors.

This No. 9A hitch tows a 694-AN planter behind a BW disk. This was one of several minimum-tillage combinations offered by John Deere in the '60s. Hitches were available to pull planters behind drawn or integral field cultivators. The 10 wheel-track hitch permitted planting four rows in wheel tracks in freshly plowed ground. Sales were low on all these combinations.

Drawn tandem disks accounted for most industry disk shipments between 1960 and 1990. In 1975, shipments of mounted tandem disks were about one-third those of drawn tandem disks. Shipments of offset disks were about one-fourth those of drawn tandem disks.

John Deere introduced a new line of nine Level-Action disks in 1973 and followed with an updated line five years later. These were called double offset, with the front gangs overlapping to take out the middles. They came in three families, based on weight per blade. The 100 integral and 110 drawn matched utility tractors and had 40 to 75 pounds per blade.

The 200 series provided the most sales, with the 210 rigid and 220 folding each having sales of more than 3,000 units in 1974. The flexible 220 had a novel system that folded the left gangs over the right for transport. The 200 series provided 60 to 118 pounds per blade, and most had 9-inch spacings. These 11- to 26-foot disks were matched to row-crop tractors. The 235 was the most popular model in the updated disks.

The 300 series disks provided 120 to 212 pounds per blade, and were spaced 9 or 11 inches apart. The most novel disk in this series was the 360 Swinger, designed for 4-wheel-drive tractor power. For transport, the operator could swing the left gangs behind the right gangs without leaving the tractor seat.

In 1986, disks and other tillage tools were transferred from the Plow & Planter Works in Moline to the Des Moines Works. In 1989 Des Moines introduced a new line of tandem disks, the 600 series. The 630 accounts for most sales and is offered in rigid and folding units. Frames are simple, with fixed-angle gangs. Folding models have hinges in the disk gangs, rather than extra framing. Widths up to 32 feet are available. Disk spacing and weight per blade are similar to the previous line. A cushioned disk middlebreaker is used to cut out the center ridge.

The 340 offset disk was popular in the Mississippi delta and other parts of the country where farmers liked this type for penetrating difficult soils. Other narrower models of offset disks were offered for use in orchards.

Here's the economical 620 integral disk on a 2550 tractor. It provides excellent maneuverability with utility tractors and produces a good seedbed.

Here's the 235 flexible 3-section folding disk working plowed ground to prepare a seedbed. The heavy frame provided durability and furnished some of the 100-pound weight per blade. Its wide swath and penetration required tractors like the 4640 or 4840.

This 630 narrow-fold disk was the favorite in 1989-90. The center section is similar to the 630 rigid, with added wings hinged at the gang frames. The hydraulic fore/aft leveling shown is an option.

The 335 was the disk of choice for deep penetration and high productivity behind 4-wheel-drive tractors. Blade spacing was 9 or 11 inches. Cone disk blades were available as well as curved blades.

Tillage

Field Cultivators

The field cultivator was well established in the market in 1960. However, it did not have the long tradition of having been popular in horse-drawn days. Its original main use was in wheat country for seedbed preparation and for weed control in summer fallow. It was used some in the Midwest for pasture renovation in the '40s. Its narrow teeth penetrated sod and hard soil. Between 1960 and 1990 its popularity spread for spring seedbed preparation in corn and soybean ground. Over the past three decades it has become a more rugged tool with better ability to shed trash.

Field cultivators and rotary cutters are the only two John Deere tools that were well developed in 1960 but had higher sales in 1990. All other major tillage tools suffered a serious loss of sales in this period. The CC-A field cultivator, made at the Horicon Works in Wisconsin, was the popular model in the early '60s. It was replaced by the C-10 and C-11 integral models and the C-20 and C-21 drawn field cultivators. The excellent 3-point hitch of the New Generation tractors from Waterloo made integral field cultivators more popular than drawn models. In November 1966 field cultivators were transferred from the Horicon Works to the Des Moines Works.

The 1000 drawn and 1100 integral field cultivators gained popularity in the '70s, with sales reaching 5,322 units in 1979. With more power available, drawn models passed integral models in sales during the later half of the '70s. The 1000 was available up to 60 feet in the 5-section folding model. These models were replaced in the '80s by the 1010 and the current 960.

Here's the well-proven CC-A drawn field cultivator working in a Kansas wheat field. It featured a rockshaft lift and down-pressure springs similar to those used on grain drills. Each shank was protected by a shear cotter pin that also served as an adjustment for pitch of the teeth. Each gang carried two shanks for 9-inch spacing or three shanks for 6-inch spacing.

This 960 field cultivator features the S-tine option. It is available in the traditional 6-inch spacing or the closer 4.5-inch spacing. Residue handling has been improved on the 960 series to provide an erosion-resistant surface. C-shank integral models are as wide as 25 feet and drawn models are as wide as 45 feet. Field cultivators continue to increase in acceptance for chemical incorporation and seedbed preparation.

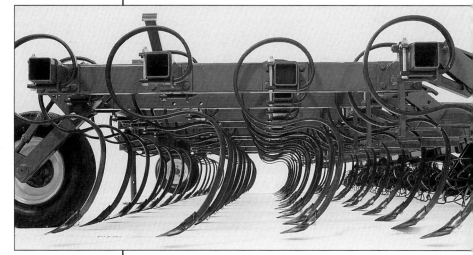

This 1000 drawn folding field cultivator is leaving rain-catching furrows. This unit has the Des Moines-designed shank attachment and spring reset. It is similar to the heavier unit used on chisel plows.

This 1010 drawn field cultivator in North Dakota shows the ability of the unit to work in trash, leaving a rough, erosion-resistant surface.

Other Seedbed Tools

Tillage tools for seedbed preparation vary widely with crops, soils, and regional exposure and preferences. Some have short lives in the market and others continue for long periods.

The F100H field conditioner was a popular tool in the late '60s and early '70s. More than 1,000 units were sold annually in 1968-70. Tine-tillage tools can be ranked by decreasing depth of penetration as V-rippers, chisel plows, field cultivators, field conditioners, spring-tooth harrows and spike-tooth harrows. Width of spacing between shanks decreases with decreasing penetration. Field conditioners provide better penetration than spring-tooth harrows, at less cost than field cultivators.

Corrugated and smooth rollers were used with horses. Planter Works engineers improved on this design in the '50s with the addition of two rows of spring teeth between the two rollers for tractor-drawn models. The roller harrow has kept a rather consistent following and was still in the line in 1990. Sales of the F925 and F950 models exceeded 1,000 units in 1968-76. Roller harrows are effective in crushing clods found in fields that have crusted after primary tillage. The front rollers crush the surface clods and the rear rollers crush those brought up by the spring teeth, leaving a fine seedbed.

The 722 mulch finisher was introduced in 1984, and sales have continued to increase with its 724 successor. It is similar in principle to the mulch tiller but is based on a field cultivator instead of a chisel plow. The lighter mulch finisher is good for chemical incorporation and producing a finished seedbed. A row of 18-inch disks is followed by three rows of cultivator shanks providing 8-inch spacing. The finish is achieved with either a tine-tooth or spike-tooth harrow attachment. The disk is fixed at 9 degrees to provide good stalk chopping and minimum draft.

The F100H field conditioner featured three rows of closely spaced spring teeth to produce a weed-free finished seedbed in light trash.

An F950 roller harrow is shown with optional crowfoot rollers. The crusted soil is adequately pulverized without engaging the spring teeth.

This 722 mulch finisher provides the action of a disk to cut residue, a field cultivator to give tillage penetration, and a harrow to produce a finished seedbed.

Here's a 30-foot flexible-fold 724 mulch finisher in transport behind a 4-wheel-drive tractor. Models vary from a 13-foot rigid to the one shown.

Miscellaneous Tillage Tools

Each of the following tools had annual sales greater than 2,000 units at some time in the '60s, but was made obsolete by more productive alternatives. Toolbars, for example, permitted farmers to develop their own tillage and planting tools, but there are now better factory-made choices. Herbicides have reduced the need for rod weeders in summerfallow. Harrows have been combined with other tillage tools.

This 500E toolbar has an added 2 1/4" square bar to mount double-coil spring standards with chisel points. Many tillage and planting options were available. The toolbar concept is built into many machines today at the factory.

Those who graduated from a one-room country school before their father bought a tractor are saddened by the passing of the spike-tooth harrow. That was the tool that marked their promotion from boyhood to being a farmer just like Dad. There are many reasons why the harrow was used for this important milestone of driving a team of horses alone in the field. The first may have been the boy's safety. Few teams ran away with a harrow and being run over by one was unlikely. The second may have been the horses' well-being. Active boys had a better memory for resting the team if they walked in soft ground behind the implement.

The harrow was a simple tool that required no lubrication and no adjustment during operation or for going back to the barn. If it caught trash, the boy could raise and clear it as it went along. Driving the horses required minimum skills. The harrowing could be done in lands or back and forth. There was no harm done if there was too much overlapping, and little harm by skipping a foot or two. Finally, the 10-foot width covered enough land so the boy could see that he had really done something, even with horses.

John Deere eliminated the loose-tooth problem on their spike-tooth harrows by resistance-welding the spikes to the bars. More than 30,000 sections were sold each year in 1960-67, with good sales continuing through the mid '70s. Although wheel carts could handle seven sections, this did not make good use of tractor power. Recent harrows are attachments to other tillage tools.

Here's a wheel-carried spring-tooth harrow breaking a crust and killing weeds. More than 20,000 sections were sold annually in 1960-1966.

This 500 series rod weeder is in summerfallow near Regina, Saskatchewan. Herbicides have reduced the need for weed control, so only a few manufacturers offer rod weeders as an attachment for chisel plows.

Planting and Crop Care

Grain Drills

Farmers a hundred years ago broadcast small grain and covered it by harrowing or disking. Well into the '40s, farmers were still satisfied to broadcast oats with an endgate seeder. There was plenty of moisture for germination following the spring sowing. But by the '30s, the horse-drawn Van Brunt grain drill made by John Deere in Horicon, Wisconsin, had an enviable reputation for seeding wheat, barley and rye. Drilling covers the seed immediately, protecting it from birds and movement by wind or water.

With the optional grass seeding attachment, drills are also used to sow grass or legumes for pastures or hay. This is usually done with oats or wheat as a nurse crop to keep weeds out of the slow-growing young crop. Drills were used in the '30s to plant soybeans for hay. As growers switched to harvesting soybeans for seed, they also switched to row-crop planters. However, the trend is now back to drilling. Rice is another crop sowed in different ways in the U.S. In Arkansas it is seeded dry with a drill and later flooded.

The fluted feedcup is used on most drills because it is easy to adjust for a variety of seed sizes and shapes. Some prefer the optional double-run meter for accuracy at low seed rates. The grass seed meter is a miniature fluted feedcup. A variety of fertilizer meters have been used over the years. However, granular fertilizer was required before any of them gave predictable rates.

The design of drills varies more with working conditions than with the crop seeded. About three-fifths of John Deere drill sales between 1960 and 1990 were for end-wheel drills. This is the preferred design on moderate-size farms of the Corn Belt and the East, where moisture is usually not a limiting factor. Most are used as single units, but 2- and 3-drill hitches have been available. Single-disk openers are standard, but double-disk openers offer better seed placement if the ground is not hard.

Here's the popular B-B plain grain drill with optional fertilizer attachment. It brought hopper capacity of 1.5 bushels per foot to John Deere drills.

The FB-B fertilizer grain drill was popular in the Corn Belt and elsewhere. It is shown with the optional grass seeding attachment, which dropped grass seed after the grain was covered.

The 8000 series end-wheel drills were introduced in the mid '70s. This 8250 fertilizer drill provided 25% more hopper capacity than the FB-B. A 2-position divider allowed the user to change the ratio of grain and fertilizer capacity.

Here's a drill liked by soybean growers. The 20-foot-wide 520 integral drill features double-disk openers. Soybeans grown in 7- or 10-inch rows soon shade out competing weeds.

This 450 series end-wheel drill was introduced in 1990. Hopper capacity has been increased to 3.4 bushels per foot, with a 3-position divider for grain and fertilizer. A new 3-section front-folding drill covers 36 feet.

Planting and Crop Care

The 500 series integral drills were introduced in 1982. Their special appeal has been for soybeans, but a variety of openers make them also suitable for other crops.

The drier areas of the U.S. and Canada have required different drills to improve germination and reduce the cost of putting in the crop. One method that was popular in the '60s but declined in the '70s was the plow-press drill. It provided a clean field that was tilled and seeded in the same pass.

The most popular press-wheel drill in the '60s and early '70s was the LZ-B lister drill with two ranks of staggered hoe openers. Press-wheel drills are readily hitched side-by-side in widths up to 60 feet. This feature has made the 9000 series, introduced in the late '70s, attractive on large farms. Sales of disk openers for press-wheel drills almost matched those of hoe openers in the '80s. The disk openers were available in 6- and 7-inch spacing, while hoe openers, for drier soils, had 7-, 10- or 12-inch spacing.

John Deere made two special drills designed to renovate pastures. The GL grassland drill sliced through the tough soils of rangeland, placing fertilizer, small grain and grass seed. Annual sales of about 100 units were made in the late '50s and in the '60s. The later Powr-Till seeder was sold in more states, with sales reaching 483 units in 1977. Farmers were reluctant to invest much money for a machine not used on a cash crop.

The 750 series no-till drills, introduced in 1989, were also designed to operate in a difficult environment but are used to drill small grains or soybeans. Each 18-inch disk is angled at 7 degrees to cut through trash and tough sod. An adjacent gauge wheel controls seeding depth, while an angled press wheel provides good seed-soil contact. The versatility of this drill has resulted in excellent sales.

Grain drill sales have been rather consistent in the 1960-90 period relative to other implements. Annual sales of John Deere drills exceeded 16,000 units in 1966-67, with the B-B providing more than 4,000 of these sales.

Here's the Tru-Vee opener, pioneered and proven on planters, applied to double-disk openers for grain drills. This is a popular option for soybeans, peas and edible beans.

A PD plow-press drill follows a packer, behind an F660-H plow pulled by a 4010 standard tractor in Kansas. The once-over operation placed the seed in the moisture of the freshly turned soil.

This 5-unit LZ press-wheel drill has wide-spaced openers. Although shown here in a trash-free field, it has press-wheel diameter large enough to roll easily over clumps of trash while providing good seed-to-soil contact.

This 4-unit 9350 press-wheel drill provides high productivity. It features a high-capacity combination grain and fertilizer box. Its disk openers provide minimum soil disturbance. The wheels for endwise transport are shown.

Here's a 9450 press-wheel drill with hoe openers, working in a field without any tillage operation after the previous crop was combined. Row markers are needed to locate the next pass through the field.

Planting and Crop Care

Various models of drills featuring a single hopper, central metering and air delivery were sold in the '80s. These wide machines have not matched the unit sales of their press-wheel drill alternative. Examples of air seeders are shown on pages 61 and 195.

Fertilizer distributors met a real need in the '50s and '60s. Annual sales of the LF drawn fertilizer distributor were more than 3,000 units in 1960-65. The Propel-R-Feed could be adjusted as low as 20 pounds per acre for fertilizer or as high as 10,000 pounds per acre for lime. The LF primarily spread before planting but was also used to sidedress corn. An integral MLF fertilizer distributor was also sold but was less popular due to its limited carrying capacity. Sales of fertilizer distributors declined as commercial application became more popular.

Here's a GL-A grassland drill working in a tough sod pasture. A notched coulter cut trash ahead of a hoe opener that placed fertilizer at the bottom. Closing knives added layers of soil over the fertilizer and between small-grain seed and the grass or legume seed.

The basic design of the fluted feed has stood the test of time. The cast-iron meter has given way to one made of modern glass-reinforced plastic. Seed rates can be adjusted over a wide range, with various seeds. Seed is not damaged and does not leak out; the meter is easy to clean.

JOHN DEERE EQUIPMENT SALES U.S. AND CANADA IN 1975

Farm tractors	50,522
Grain drills	12,636
Planters	9,383*
Unit planter units	23,382
Vegetable planters	326
Sprayers	938
Hi-Cycle sprayers	174
Rotary hoes	9,840
Front-mounted cultivators	999
Rear-mounted cultivators	6,712
Thinners	356

* Limited production of the new Max-Emerge planters reduced sales. Planter sales in 1976 were 16,546.

This 251 Powr-Till seeder works in dry or wet conditions. Its powered cutter wheels make a narrow furrow for small grain and grass or legume seed. It saves time, fuel and erosion, compared with completely tilling the land before seeding.

The popularity of the 752 no-till drill comes from its ability to work with different crops in a variety of soil and trash conditions. It helps the farmer meet increasingly strict erosion regulations. The two ranks of openers are 4 feet apart, with 2 feet of ground clearance for ample trash-handling capacity.

The LF fertilizer distributor featured a box height of less than 3 feet for easy filling. The optional grass seeding attachment is also shown.

Planters

The top four cash crops in the U.S. are corn, soybeans, wheat and cotton. The planter is a crucial tool for each of these except wheat. Planters also plant sorghum, edible beans, peas, peanuts, sunflower and pelleted sugar-beets. The following nine states have one-half of the planters on U.S. farms. Listed in descending order, they are Iowa, Illinois, Minnesota, North Carolina, Wisconsin, Indiana, Ohio, Texas and Mississippi. There is about one planter for each four tractors in the U.S. The concentration of planters is much lower than this in the wheat states.

There was a period when most corn was check-rowed with horse-drawn planters, to permit cross cultivation for better weed control. The operator had to stop and move the check wire at each end, but the horses needed a rest anyway. By 1960, however, although checking was still available on planters, the practice was doomed because of lost productivity. The tractor operator not only wasted time at the ends setting the check wire, but checking reduced top planting speeds from 7 to 5 miles per hour. By then granular-herbicide attachments were available to reduce the weed problem. As plant populations increased with better hybrids and more fertilizer use, researchers and farmers learned that drilled corn gave higher yields than checked corn. Traditions die slowly, so as an intermediate step many farmers hill-dropped corn. That practice is gone for corn but remains in some areas for cotton.

The other big change in planting corn and soybeans is in row spacing. In 1960 there was little thought of rows narrower than the 38 or 40 inches that had been set earlier for horse-drawn cultivators. By 1967 John Deere was offering narrow-row (30") planters in 6- and 8-row sizes. Deere advertising used caution in the statement, "You can switch to 36- or 38-inch rows now, and then go down to 30 inches in another year or two." As the '80s began, about one-half of all planters sold were for narrow rows. In 1990, narrow-row planters had more than two-thirds of the market. With the coming of the 9960 cotton picker in 1990, cotton can now be planted and picked in narrow rows.

Here's a 494-A planter working in a well-prepared field. It has attachments for dry fertilizer, herbicides and insecticides. It is shown as a wide-row planter but could be adjusted in 2-inch increments down to 30-inch rows.

This 1240 planter has the popular new option, a finger pickup that eliminated the inaccuracies of the plate. It also permitted planting a greater variety of sizes and shapes of seeds. Flexible rubber-covered press wheels are also shown.

The 1300 planter had single hoppers for fertilizer, seed, herbicide and insecticide for each three rows. It had walkways similar to grain drills for access to fill the hoppers. A spring-tooth harrow attachment is shown in front of the planter.

Here's the recent best seller, the drawn 6-row N 7200 MaxEmerge 2 planter. It was introduced in 1986, with a vacuum meter for improved planting accuracy.

This 8-row narrow 7000 Max-Emerge planter is shown working in a well-prepared seedbed with corn residue. Also available were Flex-Fold planters in 8-row W, 12-row N, 16-row N, 18-row N and 24-row N sizes.

Planting and Crop Care

In 1960, 2-row planters had one-fourth of the market. In 1990, the best seller was the drawn 6-row narrow planter. However, almost three-fourths of the rows sold were in 8-, 12- or 16-row planters.

Integral planters have been popular for cotton, as they make it easier to plant all the way to the end without turn rows. Integral planters provided less than one-tenth of sales in 1960, but their share expanded with the introduction of the 7100 planter in 1975. In 1990, integral planters captured one-third of the market. Drawn planters have always been more popular in corn and soybeans.

Drawn planters offer a choice of no fertilizer, liquid fertilizer or dry fertilizer. The use of liquid has grown since 1960, when only one-tenth of fertilizer attachments were liquid. Applying fertilizer with the planter declined with more use of commercial applicators, but has again increased with higher fertilizer prices and concern about runoff. Fertilizer attachments are not available on integral units due to weight.

John Deere introduced plateless planting on the 1240 planter with the finger-pickup meter. Almost two-thirds of 1240 customers chose this option. It continued as the meter of choice on the Max-Emerge line in 1975. A further step in accuracy came with the MaxEmerge 2 planters, featuring the vacuum meter. The customer now has a choice of the vacuum meter, finger pickup, feedcup, or traditional plate. The vegetable planter uses an inclined-plate meter.

Sales of planters were quite variable between 1960 and 1990. They peaked in 1966 with 24,154 planters. The 494A planter had annual sales exceeding 10,000 units in 1964-67. Sales declined in the early '70s but recovered to 20,721 of the 7000 and 7100 planters in 1979. The Tru-Vee opener made the Max-Emerge planters popular. In the '80s the trend was to more conservation planters.

Unit planters were very popular in the '70s for special applications. More than 15,000 Flexi-Planter units were sold annually in 1973-79.

Here's the integral 7300 MaxEmerge 2 planter, introduced in 1987. It featured the new vacuum meter for cotton and many other crops. Tests by John Deere showed planting accuracy in cotton improved 26%, relative to other meters.

The 7100 Max-Emerge planter, shown here in transport, brought modern planter design to integral units in 1975. It featured the new Tru-Vee openers and the proven finger-pickup meters. Sizes varied from the rigid 4-row W to the vertical-fold 12-row N.

This 484 integral planter was the choice of the large acreage cotton farmer in the '60s. It was about the only implement in the '60s that still used a second operator. He watched to see that all four units were planting the fuzzy cotton seed. Seed has greatly improved since. In the '80s farmers used monitors for determining that all rows were being planted correctly.

The vacuum seed meter has improved seed-spacing accuracy with many crops. It is a simple design, with few moving parts. Seed disks can be changed without tools. The backward-curved seed tube delivers the seed with minimum bounce and roll in the trench.

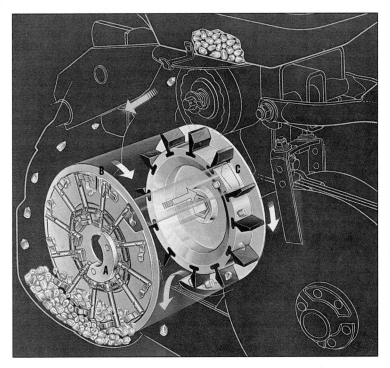

This finger-pickup meter made plateless planting practical. It showed its benefit most in corn, because it allowed accurate planting of irregular seed sizes and shapes. It lowered the cost of hybrid corn seed, as ungraded seed could now be used.

John Deere Inventiveness begins on the farm... never leaves the farm...

Scores of John Deere Engineers have farm backgrounds

From the time his fist was big enough to fit a wrench he helped Dad adjust and service equipment. As he grew, he helped improvise ways to make chores a little easier and do jobs a little better around the farm. He is still doing that—as a Product Engineer at John Deere.

A big percentage of the men engaged in John Deere design and development have a farm background. A lot of their time is still spent on the farm. That's where many ideas originate, are developed and tested, and are finally approved as new John Deere "firsts" for "The Long Green Line."

Roots-in-the-soil engineering is another reason why the John Deere franchise is the most valued in the industry.

JOHN DEERE

Here's the economical 25-B unit planter for cotton, corn and other crops. Row units could be attached to 1.75- or 2.25-inch square tool bars. More than 4,000 units were sold each year from 1960 through 1967.

The simple, functional design of the 71 Flexi-Planter made it the all-time favorite unit planter. Down-pressure on the double-disk openers was easily adjusted without wrenches. Seed plates were available for a wide variety of crops. Sprockets stored in the unit provided 13 seed spacings. Flexi-Planters originated at the John Deere Wagon Works in the '50s.

This 51 unit planter was the choice of the cotton farmer who wanted a conventional unit planter in the '70s. It featured a runner opener. A variety of spacings were available for drilling or hill dropping.

Here's the 216 potato planter planting two rows. The planter is forming ridges as it covers the potatoes. The second operator watched to see that the picker wheels continued to plant this difficult crop.

The 33 vegetable planter has met the unique needs of the vegetable farmer since the late '60s. The compact unit has a short runner opener between the front drive wheel and the rear press wheel. Well-prepared seedbeds permit planting depths of .2 to 1.7 inches. The inclined-plate seed meter is located low for accurate seed-spacings of .75 to 6.5 inches.

Sprayers

Sprayers were an important farm tool in 1960-90 to control weeds and insects in various crops. Effective, economical farm chemicals for field crops were developed after World War II. One of the early selective herbicides, still in use today, is 2,4-D. It controlled broadleaf weeds in corn and other grass-type crops. Pre-emergence herbicides were developed to prevent weeds from coming up in corn rows. Systemic chemicals that killed all vegetation, but did not sterilize the soil, were developed for planting in sod. Chemicals continue to be created that are better targeted for their end use.

Insecticides have made similar contributions to increased crop yields and quality. Boll weevil or other pest control is necessary every year in cotton. Corn borers require chemical control some years in some areas for corn.

John Deere has offered a variety of equipment to match a variety of needs. The 3-point-hitch 25A sprayer was the best seller in the late '60s and early '70s. More than 2,000 units were sold in 1966 and 1967. It was a simple, economical unit that offered a choice of six different PTO-powered pumps. Farmers wishing a wider boom and more tank capacity chose a pull-type sprayer. More than 1,000 of these sprayers were sold annually in 1963-67. Another choice was the engine-driven skid-mounted sprayer. These units fit in pickup trucks for spraying hayfields and pastures. Saddle-mounted tanks have been provided on John Deere tractors to supply sprayers on several tillage and planting tools.

John Deere's biggest contribution to chemical application has been the self-propelled Hi-Cycle sprayer. Introduced in 1962, it had annual sales exceeding 500 units six times in the '60s and '70s. It provides the high clearance regularly required in cotton and sometimes in corn. It provides spray booms 47 or 60 feet wide, and a 320-gallon tank to match. Recent 6000 Hi-Cycle sprayers have an air-conditioned cab as standard equipment.

This 25A sprayer is covering six rows of cotton. An 8-row boom was also available. Translucent fiberglass tanks held 100 or 150 gallons. It also could be used as a pre-emergence sprayer with drawn planters.

Here's a 31 trailer sprayer spraying an alfalfa field. The trailer had crop clearance similar to row-crop tractors, and adjustable tread for 24- to 40-inch rows. The 200-gallon fiberglass tank was corrosion resistant and easy to clean.

This 600 Hi-Cycle sprayer shows its 60-inch crop clearance, while spraying cotton in Mississippi. It featured a water-cooled engine with a choice of gas or diesel. The low-slung engine and 200-gallon tank provided good stability. The sprayer was useful in cotton for defoliation as well as insect control.

The cab on this 6000 Hi-Cycle sprayer provides the operator with the comfort he enjoys on his John Deere tractor, cotton picker and combine. The 62-hp diesel provides enough power to spray at 3 to 14 mph in most field conditions. Crop clearance is a generous 70 inches. Wheel treads are adjustable to fit most common row spacings between 20 and 60 inches.

Rotary Hoes

John Deere had earned a good reputation for horse-drawn and tractor-drawn rotary hoes with their cold-forged steel tine wheels. Rotary hoe function took another leap in 1960 with the added speed of the more powerful New Generation tractors. The speed helped break crusted soil and flip out small weeds.

The 14 series rotary hoe was the best seller in the '60s, with more than 30,000 sections (42" width, for one row) sold annually in 1960-67. The 4-row 14-foot integral rotary hoe was the favorite. Cable-drawn units from two to eight sections were available, as well as hydraulic-cart drawn hoes.

The next and biggest innovation by John Deere was the 400 series rotary hoes, sold in the '70s and '80s. A spring-loaded arm held down each pair of offset tandem wheels, mounted on a walking beam. This provided flexibility to fit ground contours and ride over obstructions at high speed without damage. The functional benefits were so obvious that customers bought more than 8,000 hoes each year from 1973 through 1979. The 15-foot size for four wide or six narrow rows was the best seller, with more than 5,000 units sold in 1974-76. Rigid-frame integral hoes handle six to 12 narrow rows. Integral folding hoes fit eight to 16 narrow rows.

Thinners

The John Deere thinners sold in the '70s and early '80s used advanced technology to replace the hand hoe for thinning crops. The primary crop for the thinner was sugarbeets, because the fine seeds were difficult to meter at low rates. The thinner was also used on vegetables. It was used on some cotton to compensate for excess seeding, done as insurance against poor germination. Each row unit has a feeler probe and a knife unit almost as wide as the desired plant spacing. As soon as the probe feels a plant the knife snaps across the row, removing plants ahead of the one sensed. The knife snaps back at the next plant contact.

Here's a hitch-mounted four-section 14 rotary hoe. The units are flexible in the field but become rigid when raised for transport.

The flexibility of this 400 rotary hoe is being used to fit the contour of bedded crops. It is equally effective in climbing over rocks and stumps to avoid tooth damage. Down-pressure of more than 30 pounds per wheel ensures penetration in crusted soils.

This 400 folding rotary hoe is shown in the almost folded position. The operator can switch it from full operating width to half-width transport size without leaving his seat. Four sizes, from 21- to 41-foot, are available.

This 31-foot rigid 400 rotary hoe is kicking the weeds out of soybeans. A field is completed in a short time when covering 12 rows at high speed.

This integral 6-row 200 series thinner is thinning sugarbeets. Versions of 2-, 4-, 6- and 8-row were offered. Row spacing could be adjusted on the tool bar from 12 to 48 inches. Plant spacings of 4 to 12 inches were available.

Cultivators

Corn, cotton and some other crops were first planted in rows so weeds could be controlled by horse-drawn cultivators. John Deere pioneered power lift for tractor-mounted cultivators in the '30s, and two important innovations in the '40s. The Quik-Tatch system of mounting cultivators to the front of tractors was introduced in 1947. Semi-pneumatic caster wheels were added to each rig for better gauging on uneven ground in 1949. Deere rigs were known for their parallel lift and lack of wandering when the tractor had to dodge during cross cultivation of checked corn.

In the '40s and '50s, farmers knew that the only way you could get good weed control was to set the inner shovels close to the row. This required a good view of the row to avoid cutting roots or plowing out the corn. The only place for the cultivator was in front of the operator. More than nine-tenths of John Deere cultivators sold in 1960 were front mounted, but that was soon to change.

Deere introduced the first granular-herbicide attachment specifically designed for planters in 1959. This permitted farmers to set inner cultivator shovels farther from the row and still get good weed control. As long as only two or four rows were up front, each row could be scanned for plugging with trash. The New Generation tractors encouraged the use of wider cultivators, too wide to continuously scan for plugs. But wider spacing of the inner shovels reduced plugging and tolerated some wandering at the higher travel speeds of the new tractors.

These trends, plus lower cost and easier mounting, led farmers to switch rapidly to rear cultivators mounted on the 3-point hitch. By 1963, the sales of hitch-mounted cultivators matched those of front-mounted models. In 1966 this preference let hitch-mounted cultivators capture more than three-fourths of the market. Nine-tenths of Deere cultivators were hitch mounted in 1973, and that ratio held for the next decade.

The 4-row front-mounted 40 cultivator was clearly the favorite in 1960. It was easy to mount and provided a good view of all four rows.

The T2, like its predecessor 20 cultivator, fit only tricycle tractors. Tractor tracks were removed by the rear gang mounted on the 3-point hitch.

Here's the AT-4 cultivating four rows of corn. The AT series front-mounted cultivators were designed with wide pivots to fit all row crop tractors, with either tricycle or adjustable-tread front axles. It is shown here with the 411-B side-dressing attachment.

The 725 front-mount cultivator was introduced in the '80s for the farmer who needs better steering for sidehills and working on the contour. Models up to 8-row wide or 12-row narrow are in the line. The wider units fold forward easily for transport.

The RG40 hitch-mounted cultivator took the market away from front-mounted cultivators. Rigs mounted on a 4" square tube provided the same flexibility as their front-mounted predecessors. Models were available for four to eight rows. Row spacing could be adjusted for 28 to 40 inches.

Planting and Crop Care

Two other factors influenced the adoption of rear or hitch-mounted cultivators. Dwindling sales of mounted corn pickers allowed Dubuque to drop tricycle tractors in 1965; Waterloo dropped them in 1977. Sales of tricycle tractors had been quite low since the end of the 20 series tractors in 1972. Wide-front-axle tractors were not as compatible with front-mounted cultivators as the tricycle design. The second factor was the ability of row shields to encounter trash without plugging. Recent conventional shields are suspended to swing back when trash comes along. Even better, rolling shields ride over residue while protecting small plants from soil thrown by the shovels.

The demand for the ability to tolerate trash and hard ground conditions continues to grow with the move to reduced tillage. Different models of the 800 series cultivators introduced in the early '80s met these needs. Starting at the light end is the S-tine rig for the 825 rigid and 845 folding cultivators. S-tines operate up to 7 mph for shallow cultivation. Their vibration helps them shatter clods and shed trash. Most buyers favor the single rig with three quick-return spring-trip shanks. These shanks are a smaller version of the field-proven chisel plow and field cultivator shanks.

The 875 minimum-till cultivator is similar in appearance but is built more rugged and with more trash clearance. Down-pressure springs hold the field cultivator shanks down for penetration in hard soils. Optional disk hillers slice residue next to the row. The final step up is the 885 no-till/ridge-till cultivator. Spring-reset chisel plow standards are one of the options that allow the farmer to tailor this cultivator to his practices.

Cultivator sales peaked in 1966 with 13,245 units. More than 7,000 of the RG4 cultivators were sold in 1964-67. Cultivator sales were more than one-half those of planters through 1975. Then they declined as chemicals took over weed control in row crops. Cultivator sales have increased since the mid '80s because their use is often more economical than chemicals and more favorable to the environment.

An RM630 hitch-mounted cultivator cultivates six rows of corn. It features rolling shields, down-pressure springs and spring-trip shanks.

This 875 minimum-till cultivator was designed to tolerate the harder soils and greater residue from conservation tillage. It has more muscle throughout and better trash clearance.

Here's the widest version of the most popular cultivator in the '80s, the 825 rigid. It is cultivating 12 narrow rows. Note the setting of the dual wheels on the tractor. The 800 series featured square rig beams and oil-impregnated bushings in the parallel rig links.

This 85 folding cultivator matches the width of the 16-row 30-inch Flex-Fold 7000 Max-Emerge planter. The frame folds to one-half its field width for transport.

This 885 no-till cultivator is cultivating four rows of soybeans in last year's corn stubble. Note the large shield to protect the crop from the dirt thrown by the wide shovel on the center chisel plow shank.

Hay and Forage

Mowers

John Deere's tractor mower reputation was established with the No. 5 caster-wheel mower. It lacked styling by Dreyfuss but was very functional and long lasting. It doubled the capacity of horse-drawn mowers by its 7-foot cut, higher ground speeds, and no need to rest in hot haymaking weather.

The introduction of the completely new No. 8 caster-wheel mower and No. 9 3-point-hitch mower in 1958 coincided with the introduction of a good 3-point hitch on the Waterloo 30 series tractors. The mowers were an immediate success with their clean styling, better breakaway capability and improved oil-bath chain drive. The caster-wheel design was preferred through 1962, until the 3-point-hitch design was proven in the field. The 10 side-mounted mower joined the line in 1960, and the 11 trail-type mower in 1961. The 10 side-mounted mower was the main user of the mid PTO, an exclusive on John Deere New Generation tractors.

The 30 series mowers, introduced in 1963, featured a simpler V-belt drive. Productivity increased with a faster knife speed of 1700 strokes per minute and as the 9-foot sicklebars became more popular. In sales volume, the easy-to-hitch 37 trail-type mower passed the venerable caster-wheel design in 1969. The final major mower redesign came in 1973 with the 3-point-hitch 350 mower and the trail-type 450 mower. These mowers featured a pitmanless drive.

Mowers are found wherever hay is grown for livestock. Half of the mowers on farms in the U.S. are in 10 states: Iowa, Wisconsin, Minnesota, Missouri, Illinois, Ohio, Pennsylvania, New York, Nebraska and Tennessee.

Mowers were the most popular hay tool in the early '60s, with 17,858 sold in 1962. Sales dropped steadily after the coming of mower/conditioners in 1967.

The No. 8 caster-wheel mower is shown with hydraulic lift. Hand lift or cable lift from the rockshaft were other options. More than 7,000 of these mowers were sold annually in 1960-62.

Here's John Deere's first 3-point-hitch mower, the No. 9. It is shown with a caster-wheel extension and complicated drive to a hay conditioner.

The 37 trail-type mower had offset wheels to balance the hitch. They helped keep the mower straight in heavy cutting and on sidehills. The 37 mower cut surprisingly neat corners. The 30 series mowers, with their faster sickle speeds, were able to mow up to 7 mph.

The 50 side-mounted mower was the best design for good visibility and maneuverability. High-speed mowing was possible with its pitmanless drive. This mower had one-eighth of the market in 1968 and 1969, but lost favor because it required too much time and effort to attach and remove.

The 3-point-hitch 350 was the favorite sicklebar mower from 1973 through 1990. Its clever, simple design features a V-belt sickle drive with a balanced head.

Here's the 260 rotary disk mower. It uses six disks, rotating at 3,030 rpm, to cut a 7-foot 10-inch swath. Rotary mowers work well in conditions that would plug sicklebar mowers, and they operate at speeds up to 8 mph. Rotary mowers are purchased from Europe. The 260 rotary disk mower was John Deere's best selling mower in 1987-90.

Hay Conditioners

John Deere helped pioneer the use of hay conditioners by introducing a crimper in 1955. Adoption of this new tool was slow at first because conditioning added an operation in the busy hay making season. But dairy farmers soon recognized that conditioning saved leaves and made much better alfalfa hay because quicker drying meant less exposure to sun and rain. Hitches were soon developed to pull conditioners behind mowers, eliminating the extra trip through the field. Thus, conditioning was well accepted in the '60s.

The No. 1 hay conditioner used one large and one small crimping roller, meshed like gears to crimp the stems every 2 inches. Most competitive units were crushers. In 1961, John Deere gave the customer a choice with the 21 crimper or 31 crusher hay conditioners. The basic design of the two machines was similar, except that the crusher had a large rubber-covered roll and a small steel fluted roll. This design crushed the stems to improve drying, rather than kinking and splitting them. Both designs improved hay quality, especially with legumes. They were replaced in 1965 with updated models, the 22 crimper and 32 crusher.

Sales of the No. 1 crimper hay conditioner were 3,871 units in 1960. When both crimpers and crushers were offered, crimper sales provided two-thirds of hay conditioner sales. Hay conditioners as a separate tool were only popular for a decade. Mower/conditioner sales passed hay conditioner sales in their first year, 1967.

The No. 2 swath fluffer was one of the simplest and lightest tractor-drawn tools ever sold by John Deere. Its tines rotated slower than ground speed, to lift hay that had settled from rain. This exposed more of the hay to sun and air to speed drying. The fluffer could be pulled at speeds up to 15 mph. It was introduced in 1959 and had sales exceeding 2,000 units in 1961 and 1962. By the end of the '60s its popularity had faded.

The 21 hay crimper was an updated design that built on the experience gained from the first hay conditioner made by John Deere. The lower roll picked the hay up and fed it between the two cast rolls. The 21 crimper is attached to an 11 trail-type mower.

Here's a 22 crimper hay conditioner hooked behind a 3020 tractor. The nodular-iron rolls had involute teeth for a broader contact with the stems and leaves. The weight of the conditioner was used to apply the proper pressure between the rolls, while permitting 4-inch slugs to pass through without plugging.

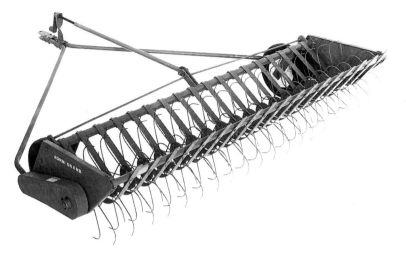

Hay that had been rained on could be lifted from the stubble and fluffed for faster drying by this No. 2 swath fluffer. Its wheels drove the fluffing reel 30% slower than ground speed.

This 31 crusher hay conditioner was the model for farmers preferring the crusher design. The machined-steel fluted lower roll pressed against a rubber-covered upper roll to crush about one-half of the stem in alternate sections.

The No. 1 hay conditioner cut curing time for alfalfa by almost one-half. Meshing cast rolls split the stems to open them up for more rapid drying. The mating rolls picked up the fresh-cut swath.

Mower/Conditioners

Farmers adopted the added operation of conditioning to haymaking in less than a decade. The merits of conditioning were rather convincing when two farmers cut their hay on the same day and the conditioned hay got put up before the rain. Rain not only bleaches the hay, but also requires that it be turned again by raking, which shatters more leaves and exposes it to more sun.

Thus, customers were ready for the next improvement, the mower/conditioner. Hitches to pull hay conditioners behind mowers were all too long, and the drive too complicated unless used with side-mounted mowers. The 483 crusher and 485 crimper mower/conditioners were adopted rapidly. The conditioning systems remained similar to their predecessors, but farmers switched their choice from crimping to crushing. The 8-foot 9-inch mower was positively fed by a finger-pickup reel for high productivity in heavy, tangled hay. Shields permitted the operator to windrow the hay to save raking, or lay it full width if it needed some drying first.

The 1209 mower/conditioner provided more than one-half of all unit sales in 1973-82. Sales exceeded 3,000 units in 1974 and 1975. Its design was a bolder step than the first mower/conditioner. This 9-foot unit had a larger pickup reel, high-flotation tires, a free-floating suspension for higher speed operation, and a clean appearance. It featured urethane rolls with interrupted cleats for higher crushing pressures and faster drying. The updated 1219 model continued to be the most popular size and model.

John Deere introduced two other mower/conditioner designs in the '80s. The 1327 used disks for cutting and an impeller with free-swinging tines. The waxy film on grass and mixed hay is scuffed off by the tines, the hood, and by rubbing against other hay. The 1460 model combines disk mowing with urethane crushing rolls.

In 1990 there were one-half as many mower/conditioners on farms as mowers. Their distribution is similar to that of mowers. In Canada, one-half of the mower/conditioners on farms are in the dairy provinces of Ontario and Quebec.

This 1209 mower/conditioner again increased productivity in haying by permitting higher travel speeds in rough hayfields. Spring-suspension floating and wide tires minimized the weight on the shoes that controlled cutting height.

Greater productivity and the ability to cut on the right or left are two important advantages of this 1600 mower/conditioner. It features the Deere exclusive urethane rolls, 110 inches long. Cutting widths are 12, 14 or 16 feet.

Here's the 483 mower/conditioner, which cut, conditioned and windrowed the hay in one pass. It used the crushing roll design proven on the 32 crusher hay conditioner.

Here's the 1470 mower/conditioner with 11-foot 6-inch swath. It combines the proven conditioning of urethane crushing rolls with a high-speed, plug-resistant disk mower.

This 1327 mower/conditioner speeds up drying by rubbing the wax off the stems of grass and mixed hay crops. The hay is cut by disks and propelled under the conditioning hood by V-shape tines. Note that the tongue is above the powershaft.

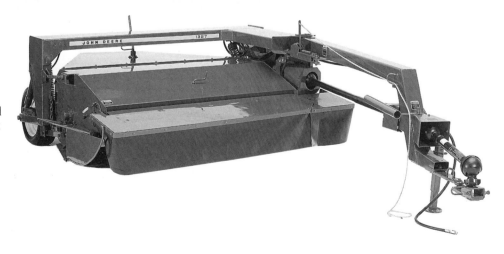

Side-Delivery Rakes

The sulky rake was the main horse-drawn rake. It was a simple tool that worked well in the low-yielding grass hays of that time. Windrow spacing could be varied with crop density by tripping the dump whenever enough crop had been gathered. A trip rope modified the rake for tractor use. Its popularity in low-rainfall areas kept it in the line, and 233 units were sold as late as 1969.

The original side-delivery rakes had cylindrical reels. The parallel-bar reel became popular in the '50s because it moved the hay to the windrow in a shorter distance, saving more leaves. In the '50s John Deere introduced a 3-point-hitch PTO rake, a 9-foot rake to match similar mowers, and rubber-ball-joint teeth, a first for side rakes. Thus, the '60s were entered with up-to-date technology as well as traditional designs.

The next step was taken in 1964 with the 640 rake. The rear wheels were offset to place the right wheel closer to the reel for better ground gauging. Universal joints between the wheels allowed both wheels to power the reel. Productivity was high, as the rake moved a 9-foot swath at speeds up to 7 mph.

The last update in the 1960-90 period came in 1988. The 652 integral rake uses hydraulic instead of PTO drive. Hydraulic drive is an option on the previously ground-driven units. Two choices are given for higher productivity. The conventional 672 can be paired with the right-hand-delivery 673 to rake 18 feet. The 700 twin rake handles up to 23 feet.

Wheel rakes, sold from the mid '60s to mid '80s, could rake up to 15 mph in light grass hays. Wheel rakes had one-sixth of the rake market at the height of their popularity.

Side rake sales peaked in 1965 at 8,541 of all models, and with more than 5,000 sales of the 640 rake in 1965-67. Rake sales declined in the '70s and '80s, with increased mower/conditioner sales. There are about three-fourths as many rakes as mowers on farms in the U.S.

Here's the 350-A integral side rake. This economical rake had good ground gauging with its two caster wheels and close coupling to the tractor. It had a positive PTO drive.

This pull-type heavy-duty 894-A side rake was the choice of large alfalfa growers in the early '60s. Both rear wheels provided positive ground drive for the parallel-bar reel.

Here's the 670 side rake. Both wheels provided positive ground drive matched to travel speed to minimize hay movement. A 5-bar reel was standard.

This ground-driven 640 side rake turns a full 9-foot mower swath. The stainless steel rubber-ball-joint teeth reduced repairs and kept broken teeth out of the hay. This was important for the life of balers and forage harvesters. It was even more important to keep metal out of cattle feed.

The 567 wheel rake is a simple design that moves the hay gently at high speeds. The optional 7-wheel 11-foot rake could be operated up to 15 mph for high productivity.

Square Balers

John Deere introduced the first wire-tie pickup baler in 1946. By 1950, it was obvious that most new buyers of pickup balers favored a twine-tie model. The 14-T was well accepted by first-time customers. In 1957 John Deere was the first to offer the bale ejector. For the larger acreage hay grower, the 214-T was the choice. The needs of the commercial hay grower were met by the 214-WS. It featured the single-twist knotter, while previous Deere balers had a wire knot at both ends of the bale.

For the custom operator and the large commercial grower, the 323-W 3-wire baler was developed. It produced large, heavy bales 16x23 inches and up to 50 inches long. The 323-W was powered by a 36-hp Deere engine. It was sold primarily in the Southwest, where hay is normally shipped long distances by truck.

The 24-T replaced the 14-T in 1962 and the 224 replaced the 214 in 1964. Both continued the popular 14x18-inch bale size of their predecessors. The 224 could produce denser bales from its 4-sided bale tension system. Rollers on the plunger kept the knife in register longer than previous designs. With the exception of 3-wire bales, all bales were light enough for a man to lift.

In 1963, John Deere introduced a different type of baler to the market. The 10 Hi-Density baler produced 10x15-inch bales with double the density of conventional bales. Trucks could haul more hay, and less space was required for storage. This density was obtained with a high plunger speed, hydraulic density control and two wires.

In the '80s, customers had a third choice in 14x18-inch balers with the 327, 337 and 347 models. Each baler used a gear- and shaft-driven knotter that eliminated the chain drive. Chain drives could break and throw the baler out of time. The two lower-cost balers had a plunger speed of 80 strokes per minute. The heavy-duty 347 used a higher speed of 93 strokes per minute for more capacity.

This 214-T twine-tie baler was similar in basic design to the 14-T but built heavier throughout. Twine was preferred if the bales were to remain on the farm.

Here's the heavy-duty 347-W baler dropping a bale on the ground. Wire was chosen for the repeated handling of hay that was shipped off the farm. Grooves were formed in all John Deere bales to keep the wire or twine from slipping off during handling. Elimination of the chain drive to the knotter solved one of the nuisance problems of earlier balers.

Here's a 14-T baler with bale case extension to guide the bales back for loading on the wagon. The economical 14-T helped many dairy and beef farmers switch from loose to baled hay for easier handling, storage and feeding.

The economical 328-T baler ended the '80s as the best seller. It is quite similar to its bigger brothers, now having their 74-inch-wide pickup. Close-spaced pickup teeth thoroughly pick up the windrow.

The 8 series balers were introduced in 1987 and included the 468 that made heavier 16x18-inch twine-tie bales.

The bale ejector made one-man haying practical for the many farmers who had no hired help. The operator could work alone in the field because the ejector threw the bales in the high-sided wagon. At the barn the half-size bales were dropped into a conventional flight elevator. A bale conveyor suspended from the barn roof received the bales and dropped them at the selected location in the haymow.

The first ejector threw straight to the rear of the baler and sometimes bales missed the wagon on turns. The No. 2 bale ejector corrected this by pivoting the ejector to follow the wagon. The 30 bale ejector brought the smoothness of hydraulic actuation. It also threw the bale by swinging the pan underneath it rather than grabbing the sides. This reduced leaf loss and permitted the operator to easily switch back to dropping the bales on the ground.

Square baler sales were about level in the '60s and peaked at 10,923 units in 1973. Sales of square balers declined with the introduction of round balers in 1975. Many dairy farmers still prefer the hay quality from square bales. The lowest cost twine-tie baler has consistently been the best seller, providing more than half of all sales in the '60s and '70s. The preference for wire-tie balers grew in the '60s and accounted for about one-fifth of all baler sales in the '70s. Bale ejectors were sold on one-fourth of balers in the '60s and on more than one-third of balers in recent years.

The 10 Hi-Density baler never caused much excitement, with sales peaking at 99 units in 1964. The 323-W 3-wire baler had even lower sales. The field hay cuber, described on pages 25 and 26, did slightly better, with sales averaging 68 units yearly for a decade.

Wisconsin, Texas, Minnesota, Pennsylvania, New York, Ohio, Missouri, Idaho and Michigan had over half of all industry sales of square balers in 1989. There are about three-fifths as many square balers on farms as mowers.

The 323-W 3-wire baler was sold primarily in the irrigated Southwest to commercial hay growers. The heavy bales were picked up mechanically and loaded on semis for transport to dairies and beef feedlots, often hundreds of miles away.

Here's the No. 5 flail pickup on a 24-T baler salvaging corn stubble. The small bales are being thrown into a 112 Chuck Wagon to be used for bedding.

This 10 Hi-Density baler is working in doubled windrows. It has the optional engine drive. The pickup was similar to other John Deere balers but the main drive was different. The dense wire-tie bales required only half the space for storage. However, the hay had to be drier when baled, to avoid spoilage.

This team consists of a 2010 tractor, a 14-T baler, a No. 1 bale ejector and a wagon with high sides and low front. A feeler tripped the ejector when a bale came through the case. An arm on either side of the bale grabbed it and sent it up the curved chute to the wagon.

Here's the 40 Hydra-Load bale ejector tossing a larger bale in the wagon. The pan-type thrower on the 40 handles bales up to 38 inches long and 80 pounds. An optional control in the tractor permits the operator to set both the direction and distance of the throw.

Stack Wagons

Beef cattle farmers have a long tradition of stacking loose hay outside and letting the cattle self-feed directly from the stack. In the more humid areas there was a considerable loss of hay quality. This was reduced by manually stacking the hay and ending with a rain-shedding top. If the cattle were not fenced properly from the stack, there was also a significant quantity loss.

In the northern Great Plains, with low rainfall, farmers eliminated the man on the stack and made huge stacks with a variety of stackers. John Deere entered this market in 1962 with the 52 hay stacker. It was a very functional unit and mounted on row-crop tractors. But it had poor acceptance in the market and was dropped.

By the '70s, farmers were insisting there must be some better way to make hay than loose stacks or manually lifting 50- to 80- pound bales. Competitors introduced two ways of mechanically making large hay packages— stack wagons and large round balers.

The Des Moines Works introduced the 200 stack wagon in 1974. A flail pickup lifted the cured windrow and blew it into the wagon. As the coarsely chopped hay filled the wagon, the operator stopped and compressed the hay by lowering the top canopy. This was done a few times with pressures up to 184 pounds per square foot to produce a dense 4-ton stack. The canopy produced a round top to shed rain and snow.

The 1.5-ton 100 stack wagon and the 6-ton 300 stack wagon followed in 1975. Stack movers were sold to match each size of stack.

More than 1,000 of the 200 stack wagons were sold in 1974 and 1975. However, farmer preference for large round bales caused stack wagons to remain in the line less than a decade. The flail pickup of the stack wagon lost more leaves than the tine pickup of round balers. Stacks frequently settled unevenly, leaving pockets to collect rain and cause spoilage. Stacks were harder to move.

Here's a 52 hay stacker building a large stack on the plains. The 12-foot basket builds stacks 24 feet high. Loader booms extend to the rear and outside the wheels to provide height and stability.

Here's a 100 stack mover loading a stack made by the 100 stack wagon. Its three 10-foot tines slid under the stack before it was hydraulically raised on caster wheels for transport.

This 100 stack wagon produces a dense 1.5-ton stack 7 feet wide, 10 feet long and 8 feet high. A conveyor chain in the bottom of the wagon is used to unload the completed stack on the ground.

This 200 stack wagon is picking up a windrow of hay to produce a 4-ton stack 8.5 feet wide, 14 feet long and 10 feet high. Note the hydraulic cylinder that pulls the canopy down to make a dense stack. Stack wagons were also used to pick up cornstalks for bedding.

The 300 stack wagon was similar to the 200 in design but made a 6-ton stack 21 feet long. Some farmers unloaded the stack in the field and moved it later with a stack mover. Others transported the stack to the site where it was to be fed before unloading.

Round Balers

John Deere entered the market with the 500 round baler in 1975. Windrowed hay was picked up with a small-diameter tine pickup and fed into a multiple flat-belt variable-size chamber. When the bale was complete, the tractor paused while a twine arm crisscrossed the bale to wrap it. Then hydraulic cylinders opened the rear of the bale chamber and discharged the bale.

The updated 410 and 510 balers introduced in 1977 had improved feeding. Introduced in the early '80s, the 430 and 530 balers featured diamond-tread belts, automatic twine wrap and the exclusive Bale-Trak monitor. The monitor helped the operator weave across the windrow to produce a cylindrical bale without looking back. Four sizes of 35 series balers were introduced in 1989. The 335 makes bales 46 inches wide, up to 51-inch diameter and 750 pounds. The 535 makes bales 61 inches wide, up to 72-inch diameter and 2,000 pounds.

Unlike most other farm equipment, round balers sold better in the '80s than in the '70s. Sales were 4,422 units in 1979. The largest size has consistently provided more than half of all sales. This is the opposite of the experience with square balers. The largest bale is popular because it reduces the number of bales that must be transported from the field. Some bales are stacked two or three high inside sheds, especially by dairy farmers. Most round bales made for beef cattle are left outside.

In 1990, John Deere offered three ways to handle round bales. The grapple attachment on a loader provides the most positive method. The integral 84 bale mover also can be used for feeding. The 100 round-bale fork is the economical solution and is mounted on a loader or the 3-point hitch.

Canada illustrates the influence of rainfall and type of livestock on the selection of balers. In 1986, Ontario and Quebec had almost half of the square balers on farms. Alberta and Saskatchewan had almost two-thirds of the round balers.

The smallest of the round baler line, the 330, has just released its bale. Forward travel is resumed after the rear half of the bale chamber is closed.

This 535 round baler has applied two turns of surface wrap on the bale. This woven mesh wrap provides better weather protection but allows the bale to breathe to cure without spoilage. The optional bale push bar, combined with the reduced time for wrapping, allows the operator to start baling again only 15 seconds after stopping for bale completion.

Three optional features are shown on this 435 round baler. The rubber converging wheels help gather in wide or wind-blown windrows. Gauge wheels on the pickup provide cleaner pickup and less tine damage in rough or undulating hayfields. The surface wrap attachment is shown on the rear of the baler.

Here's a 500 round baler operating in a rolling alfalfa field, with dairy buildings beyond it in the valley. This baler made an immediate hit with farmers, with second-year sales greater than 2,000 units.

The 410 was an economical round baler that produced smaller bales that could be picked up by loaders on utility tractors. It featured more dependable feeding than the earlier 500 baler.

Forage Harvesters

Silage making in the '30s was quite labor intensive. Horse-drawn corn binders made bundles. These were loaded on a wagon and hauled to the tractor-powered ensilage cutter at the upright concrete silo. Spoilage was expected on top and wherever there were leaks in the concrete walls.

Two developments influenced a trend to more silage. In the '50s and '60s, blue glass-lined silos were adopted rapidly by dairy farmers. These oxygen-free silos unloaded from the bottom, improving silage quality and reducing feeding labor. A different approach used for beef was the bunker silo. It offered minimum investment per ton stored and could be unloaded with a loader.

John Deere moved silage cutting to the field in 1941 with the first flywheel-type pickup forage harvester. The plate-type flywheel combined the function of cutting with blowing the material into the trailed wagon. This design was used in the No. 6 and the 12 forage harvesters of the early '60s.

Completely new cylinder-type cutterhead forage harvesters were introduced in 1966. The economical 34 was the cut-and-throw design. The high-capacity 38 separated cutting from blowing for finer cuts. The windrow pickup was simpler and fed better. The row-crop units featured positive-feeding rubber gathering belts. Recent Deere forage harvesters use segmented knives, which allow lower-cost repair of cutterhead damage.

Forage harvester sales peaked at more than 5,000 units in 1974 and 1975. Sales of the 38 and 3800 harvesters exceeded 2,000 units yearly for six years in the '70s. Forage harvester sales declined greatly in the '80s as silo construction almost stopped.

Half of recent industry forage harvester sales in the U.S. were in the dairy states of Wisconsin, New York, Minnesota and Pennsylvania. There are about one-third as many forage harvesters as square balers on U.S. farms.

The heavy-duty 38 forage harvester easily handled two rows of corn with the new row-crop header. Chopped material was augered from the cutterhead to the blower. This was the best selling size of forage harvester because it met the needs of custom operators and large acreage farmers. There are about two row-crop units sold for each three forage harvesters, with the 2-row narrow unit the most popular.

This side-mount 25 forage harvester was imported from France in the late '70s and early '80s. It was a very economical machine for the small dairy farmer who fed only corn silage.

Here's the medium-duty 3940 forage harvester with 5.5-foot windrow pickup in alfalfa. About two windrow pickups are sold for each three forage harvesters. The 7-foot pickup passed the 5.5-foot pickup in sales in 1988.

Here's the popular forage harvester of the early '60s, the No. 6. Its 6-knife cutterhead produced lengths of cut of 7/16 to 2 inches. The gathering chains and sickle of the row unit were very similar to those of their corn binder predecessors.

This cut-and-throw 34 forage harvester is using a 6-foot mowerbar in heavy alfalfa.

This heavy-duty 3970 forage harvester is a match for tractors up to 190 hp. It has .25- to .5-inch length of cut and can be fitted with a recutter screen. Harvesting heads include four row-crop units, two hay pickups and one mower/bar.

Hay and Forage

John Deere introduced the 5200 and 5400 self-propelled forage harvesters in 1972. In addition to the types of harvesting units available before, a Stalker was added. It was similar to row-crop units but was designed to salvage stalklage after corn had been combined for grain. The Stalker remained in the line for over a decade but was never well accepted. There was too much investment in time and money for the value of the crop salvaged. Corn heads that are used on combines are available for chopping ear corn. An optional kernel processor is available to crack the corn.

With the greater use of self-unloading forage wagons, forage blowers shifted from conveyors to hoppers in the '60s. The other major trend in 1960-90 was greater blowing heights. This was helped by using hoppers with rotary tables that accelerated the crop before it reached the fan. John Deere made the 55 and 60 blowers sold in the '60s. The 65 blower sold in the '70s was made by Kools. The recent 100 and 150 blowers were made by New Holland.

JOHN DEERE EQUIPMENT SALES U.S. AND CANADA IN 1975	
Farm tractors	50,522
Mowers	4,629
Mower/conditioners	4,159
Side-delivery rakes	3,756
Square balers	7,915
Round balers	788
Stack wagons	1,329
Forage harvesters	3,874
Forage blowers	707
Forage wagons	1,701
Rotary choppers	193
Flail shredders	846

The 5400 self-propelled forage harvester brought increased productivity to cutting silage. The hydrostatic rear-wheel drive option shown helped keep harvest on time, even in rainy weather. The 1510 Hi-Dump wagon reduced unloading time.

The 60C conveyor blower eliminated the right-angle gearbox. It is shown for 540-rpm operation. The small-diameter fan could be driven directly by a 1000-rpm PTO. The operator could control the unloading rate of the Chuck Wagon for speedy unloading without plugging the blower.

Here's the ultimate in forage harvester capacity and convenience— the 5830, introduced in 1986. A variety of options are offered in addition to several header units for hay or corn. The Spout-Trak control moves the spout to follow wagons around corners. The Iron-Gard metal detector protects the machine and the cattle that eat the silage.

This 55 forage blower had a spring-balanced hopper that was lowered after the wagon drove by. The conveyor chain in the bottom of the wagon had to be connected to its drive from the blower. The operator was kept busy trying to get uniform feeding from the wagon.

The 65 forage blower was a simple unit, designed to be used with self-unloading forage wagons. Cycle time at the silo was reduced by quicker setup and more uniform feeding.

Forage Wagons

Forage wagons evolved as farmers became willing to invest more in equipment to save time and labor during the crucial silo-filling season. Optimum silage production requires catching the right range of moisture as nature dries out the hay or corn crop.

Wagons of the '40s and '50s depended on a drive from the blower to unload the wagon at the rear. Some early units had a false endgate pulled to the rear by chains. Others used a heavy tarpaulin spread on the wagon floor and then wound to the rear on a roller. Although it made the wagon more complicated and expensive, the unloading design that won out was the conveyor chain. It operated at a uniform speed, but the chopped material came off in chunks, requiring considerable effort to avoid plugging the blower.

Engineers at the John Deere Spreader Works, East Moline, Illinois, introduced a better solution in the late '50s—the 110 Chuck Wagon. It had two beaters in front that broke up chunks in the silage before a cross conveyor delivered it to the forage blower hopper. The conveyor provided five speeds to match the harvested crop and the blower capacity. The Chuck Wagon performed another important job on the farm, unloading stored silage into cattle feedbunks. Grain spread on top of the load would mix with the silage. Annual sales of Chuck Wagons exceeded 2,000 units in 1962-1966.

A lower-cost alternative was introduced in 1966—the 214 and 216 forage wagons. These were high-capacity units for hauling silage from the field, but were not designed for feeding cattle. In 1977, these wagons and Chuck Wagons were replaced by the 714 and 716 forage wagons, made at the Welland Works in Ontario, Canada.

The updated 714A and 716A wagons were sold through the '80s. They feature steel construction, with epoxy-coated side sheets. A variable-speed V-belt drive provides smooth delivery from the conveyor chain. An optional hydraulic control permits the operator to unload into the blower or feedbunks without ever leaving the tractor seat.

Here's an additional use for the 115 Chuck Wagon besides hauling silage to and from the silo. At the barn, the rear tailgate of the wagon could be opened and the small bales slowly conveyed into the bale elevator.

The 716A forage wagon provides high productivity in the field and in the feedlot. It also can be unloaded from the rear for bunker silos. With high sides, it has 677 cu. ft. capacity. It is shown on the 12-ton 1075 wagon gear.

This 216 forage wagon was the right choice for the operator who had lots of crop to deliver to the silo but did not need the wagon for feeding. It was constructed of yellow pine.

Flail Machines

In the '50s, many dairy farmers became convinced that they could keep more cows per acre if they brought the pasture to the cow. This was especially true if they had no system for rotating pastures. John Deere's answer to this problem was the 15 rotary chopper. The farmer could green-chop standing hay and bring it fresh to the dairy cattle each day. Free-swinging curved knives cut the crop and conveyed it back to a cross auger that transferred it to the blower.

In 1961, three knives were added to the fan to further reduce the length of cut. The 6-foot 16A rotary chopper was added that year for higher capacity. It is one of the few John Deere products to remain in the line for three decades with minimal design changes. Some rotary choppers have been used to fill horizontal silos, but the length of cut is too long for ideal packing of the silage. Rotary choppers can be used to collect straw or cornstalks for bedding. With the auger door open, they can be used to chop stalks, weeds or light brush. More than 3,000 rotary choppers were sold in 1961.

The same basic rotor and knife design was used in the 26 flail shredder, also introduced in 1961. The 27 flail shredder, sold in the '70s and '80s, was primarily used to chop cotton stalks, cornstalks and other crop residues for insect control and faster decomposition. It is more effective in lifting fallen stalks between the rows than rotary cutters. Although a flail shredder could be used for mowing pastures, its greater number of knives makes it more expensive to maintain than a rotary cutter.

The 25A flail mower is a third machine that is based on the same principle of cutting. It is a good institutional mower because it is quiet and provides a better cut than rotary cutters. Cutting height is well controlled by a roller immediately behind the rotor. A steel and rubber shield deflects cut material down uniformly across the width of the mower.

The 15 rotary chopper and 110 Chuck Wagon were a good pair for green-chopping alfalfa at the start of the '60s. Rotary choppers provide a clean cut with minimum leaf loss.

This 6.5-foot 25A flail mower will operate up to 7 mph. It is most effective on institutional lawns and roadsides where uniform height of cut and even distribution of clippings are important. Its 100- mph-plus tip speed helps it digest twigs, cans, and other litter with ease.

This 27 flail shredder is pulverizing four rows of cotton stalks for better boll weevil control. Knives are faced with tungsten carbide for longer wear. The shredder is equally at home in six 30-inch rows of cornstalks.

Here's the venerable 16A rotary chopper, still a favorite of dairy farmers for bringing the pasture to the cow. It is a simple, rugged and durable machine that can be depended upon for daily cutting.

Harvesting, U.S. and Canada

Corn Pickers

Corn pickers were in their heyday in the '50s. The mechanized harvesting of small grains had arrived much earlier, in the days of horses and steam engines. There was less urgency then to mechanize the corn harvest, because manual productivity was relatively high. Snapping one ear of corn was much faster than cutting an equal amount of wheat. Picking corn could be done from the standing position in cool weather. A team of horses could be taught to follow the corn row and keep the wagon near the man that was picking. But all that changed with tractor power.

Mechanical corn pickers used a pair of snapping rolls that pulled the stalk down through, much like the action of a wringer on a washing machine of the '30s. The ear was snapped off when it hit the rolls. Some kernels were shelled off and lost. The ear was then conveyed back to a husking bed, consisting of paired rubber rolls that removed the shucks. A fan blew away silks and some husks as the ears fell into the elevator that conveyed them to the towed wagon.

Customers preferred mounted pickers over drawn units because vision was much better for staying on the row and getting under down stalks. Mounted pickers were more maneuverable and more mobile in soft fields because they put some of the picker's weight on the tractor drive wheels. But mounted pickers also had their faults. They were difficult to put on and take off, so the tractor could not be used for other jobs during corn-picking season. It was impractical to mount a cab on a tractor with a picker because there was no room for the door to open. Air intakes to the radiator had to be enlarged because of the trash in that vicinity.

John Deere was the first in the industry with a corn head for combines in 1955. The decline of corn pickers was well under way in 1960, but sales were still double those of corn heads. That year the 227 corn picker had sales of 5,054 units, and the 100 corn snapper had sales of 663 units. Corn head sales passed those of corn pickers in 1963, and tripled corn picker sales by 1965.

The 2-row tractor-mounted 227 picker was a favorite in the Corn Belt. John Deere was first on the market in 1958 with a sheller attachment. The 50 sheller, shown here, produced clean shelled corn similar to combined corn. On wagon gear, the barge box was beginning to take the place of flare boxes for grain.

The mounted 237 corn picker was a worthy successor to the 227, but annual sales never reached 2,000 units. The picker was designed for speeds up to 5 mph and yields of 100 bushels per acre. Spiral fluted snapping rolls cut shelling losses in half.

In the mid '60s the side-mounted 1-row 120 corn snapper was the low-cost solution for the farmer without a self-propelled combine. It offered the maneuverability of mounted pickers and could be hitched almost as easily as drawn pickers. It snapped the ears but did not remove the husks.

Here's the drawn 300 corn husker, Deere's most popular unit in the declining market of the '70s. It featured reduced shelled corn losses by using either a 2- or 3-row 40 series corn head for the snapping unit. Some farmers with good corncribs or a livestock operation preferred to stay with ear corn. Seed corn producers also favored harvesting without shelling for better seed quality. An Iowa manufacturer was still making a 2- or 3-row drawn corn harvester in 1990, using a John Deere corn head.

Windrowers

When farmers switched from grain binders to combines, harvesting was delayed until most grain was at a storable moisture content and easier to thresh. In the spring wheat areas of Canada and the Dakotas, this was not practical. Grain ripened later in cooler weather and was subject to losses from hailstorms. Some old grain binders were converted to windrow the grain so it would dry out earlier.

These were soon replaced by drawn, PTO windrowers, designed for the job. The basic design of PTO windrowers was similar to the cutting platform of binders, with a bat reel, sicklebar and horizontal canvas draper that conveyed the grain to one side. Self-propelled windrowers followed, with greater productivity. They also opened fields without knocking down standing grain.

Sold in the '60s, the 16-foot 190 PTO windrower had a double-swath option to make 32-foot windrows. Spring wheat and barley yields are low in low-rainfall areas, so wide coverage is needed to take advantage of the capacity in large combines. The self-propelled 215 grain windrower, introduced in 1961, soon outsold PTO windrowers. Farmers recognized the potential of this machine for cutting hay, so an auger platform was offered in 1965. Draper platforms for grain had higher sales than auger platforms until the late '80s. Heavier duty auger platforms and traction units were developed for hay.

Sales of PTO windrowers held up well from 1960 to 1990. The 580 provided the highest PTO windrower sales in the early '80s. At that time, Canada accounted for three-fourths of 580 windrower sales. More than half of PTO windrowers on Canadian farms are in Saskatchewan.

Self-propelled windrowers had their highest sales in the '70s, with annual sales of the 800 windrower exceeding 2,000 units in 1972-76. Most sales of draper platforms for grain are in South Dakota, North Dakota, Colorado, Saskatchewan, Alberta and Manitoba. Most sales of auger platforms for hay are in California, Oregon, Wisconsin, Montana, Kansas, Idaho and Nebraska.

The 190 PTO windrower laid heads-in-line windrows on top of the stubble, for fast drying and easy pickup with minimum shattering. It is shown adding the second windrow to make a double windrow for better use of combine capacity. The unit was easily switched to transport position.

The 215 windrower offered the advantages of a self-propelled unit in 10- to 16-foot widths. It is shown cutting wheat with a draper platform.

Here's the popular 580 PTO windrower. It was available in widths up to 36 feet for high productivity. It was made for John Deere by MacDon Industries in Winnipeg, Manitoba.

This self-propelled 800 windrower was designed primarily for grain and had only draper platforms, which came in widths of 10 to 21 feet. Although draper platforms can be used for hay, as shown, auger platforms are more trouble-free. The 800 featured two rear caster wheels for better weight balance.

Here's a self-propelled 2320 windrower with 12-foot auger platform in hay. It was the best selling self-propelled model in grain in the '80s. Draper platforms included single-swath sizes up to 25 feet. There was also a twin-swath platform that laid 42 feet of grain in side-by-side windrows.

Pull-Type Combines

It was the small, straight-through pull-type combine that switched farmers away from grain binders and threshers in the '40s and early '50s. Most had a 5- or 6-foot cut and a full-width cylinder and straw rack. They could be operated by a 20- to 30-hp tractor. Or they could be powered by an optional engine and pulled by a smaller tractor.

The self-propelled 55 combine, introduced in the '40s, was an immediate hit in the wheat country, but had limited sales in the Corn Belt and the East. The 45 self-propelled combine came in the '50s and made real inroads on the small PTO combine market. The 30 PTO combine was a refined version of its popular predecessors. It still outsold each of the self-propelled models in 1959, but sales were slipping fast. In 1961, the 42 PTO combine was offered to Midwest farmers with only modest success. It and its 40 self-propelled combine counterpart were in the line less than a decade. Farmers with limited acreage depended on their neighbors to custom-combine their crops.

Some farmers with large acreage needed the capacity of self-propelled combines, but their low grain yields made it difficult to justify the cost. The 65 PTO combine (based on components of the 55) was introduced in the '50s to meet this need. The 96 PTO combine was introduced in 1963 to match the added power and better PTO of the New Generation tractors. It had annual sales exceeding 1,000 units in 1964-66, making it Deere's best selling large PTO combine. Sizes have continued to move up with increasing tractor power. They include the 106, 6601, 7721 and 9501 PTO combines.

In the early '80s, 90% of John Deere pull-type combines were sold in Canada. Saskatchewan has almost half of all pull-type combines on Canadian farms. Practically all large PTO combines have belt pickups for windrowed grain. During the past decade, there has been some shift toward direct cutting of standing grain in the Dakotas.

The 30 combine and its predecessors were ideal for the first-time combine owner. They were low in cost because they were tractor powered. The auger platform on the 7-foot-cut 30 combine was more durable than canvas, and fed more dependably. The 30 was the last PTO combine to have the cutting platform on the left side.

The 42 combine provided pull-type combine buyers with a 9-foot grain platform for small grain or soybeans. For the first time, they also could harvest corn with the 2-row 205 corn head.

Here's the 96 combine, the all-time favorite pull-type combine for windrowed grain. It featured a 40-inch-wide separator, and an 80-bushel grain tank with fast unloading.

This 7701 PTO combine has a wide belt pickup designed so it could pick up side-by-side windrows, to use the full capacity of this large combine.

The pull-type 9901 combine provides the windrowed-grain farmer with the advantages of the Maximizer combine at a much lower cost than comparable-capacity self-propelled combines. The operator adjusts the combine from a console in the tractor cab.

Self-Propelled Combines

John Deere made a major contribution to self-propelled combine design in 1946 with the introduction of the 55 combine. It was the first machine in the industry to center the operator on top and locate the grain tank and engine directly behind the operator. This gave better weight balance, with or without grain in the tank. It also was a cleaner, quieter location for the engine.

The 55 combine met the needs of the wheat farmer in the West, but by the '50s there was need for additional sizes of self-propelled combines. The smaller 45 combine was designed for the mass market of farmers wishing to switch from pull-type combines to self-propelled units. It used a spot-welded formed-sheet-metal frame, similar to cars. The wheat farmer was ready to move up in capacity. John Deere engineers responded by taking the proven 55 combine design and increasing the number of straw walkers from three to four. Introduced in 1958, the 95 combine had a 40-inch cylinder and up to an 18-foot platform.

The '60s opened with the introduction of Deere's smallest self-propelled unit, the 40 Hi-Lo combine. The updated Hi-Lo combines had wider grain tanks, and the top of the clean grain elevator was now concealed in the tank. In 1961, the fifth straw walker was added to a strengthened version of the basic 55 combine design. The 105 featured 14- to 22-foot platforms, a 105-hp engine and a 75-bushel grain tank.

The drives on the 55, 95 and 105 combines were similar except for size. The separator was clutched by a tightener on the flat-belt drive from the engine to the beater. The 4-speed transmission was driven by a variable-speed V-belt drive from the engine. In the mid '60s, hydrostatic ground drive became an option on these combines. The flat-belt separator drive was replaced with a double V-belt drive.

The 45 combine was Deere's best seller from 1960 through 1964, when the 95 took over that honor for the rest of the '60s.

Here's the self-propelled 40 combine cutting two rows of grain sorghum. It featured an 8- or 10-foot grain platform, a cylinder 24-5/8 inches wide, a 42-hp engine and a 35-bushel grain tank. It was Deere's lowest priced self-propelled combine, selling for about $5,000 complete with 8-foot platform.

This 55R combine is operating in a muddy rice field. Note the reverse-slope windshield that stayed clean longer, main drive shields, extended air cleaner intake, and LP-gas fuel tank.

This 45 combine is harvesting soybeans in Iowa. The 45 was the favorite combine for the first-time buyer who had corn, soybeans and wheat.

The 95 combine with 410 corn head had twice the corn harvesting capacity of corn pickers or previous combines with 2-row corn heads. The 410 was basically two side-by-side 210 corn heads. In the early '60s the heads had to be operated close to the ground to harvest all the corn. The design allowed each 2-row unit's height to be adjusted separately.

Corn harvest capacity took another step up in the late '60s with the introduction of the 6-narrow-row 635 corn head, shown on a 105 combine. The operator had an excellent view of the corn head, for keeping the unit on the rows and the gathering points at the desired height.

The '70s ushered in the New Generation combines. The 3-walker 3300 and 4-walker 4400 shared one design, and the 4-walker 6600 and 5-walker 7700 shared another. These four machines had a similar clean appearance. The engine was relocated in front of the grain tank, and the cab was offset to the left of the engine. The engine was completely enclosed, with a large front screen. A reverse-flow engine fan pulled air through this screen and discharged it through the side, beyond the radiator. The cab was integrated into the design of the combine, resulting in reduced noise and dirt for the operator.

Productivity was increased with better functional design of the separators, more power and larger grain tanks. Productivity was also increased with the new Quik-Tatch platforms and corn heads, which made it practical for the Corn Belt farmer to readily switch back and forth when harvesting soybeans and corn.

The combines were updated in 1974, with the most visible difference being the engine cooling system. A belt-driven rotating screen on the side of the combine had a vacuum arm that removed any accumulated trash from the screen.

Engineers are never satisfied that they are offering customers all the choices they might want. In the early '60s a self-propelled corn harvester was run experimentally. Its general layout was similar to a combine but much simpler. Farmers had the option of a shelling unit or a husking unit. They could even switch back and forth. The corn was conveyed to a trailing wagon.

In 1969, the Des Moines Works discussed a machine to harvest shelled corn, ear corn or corn silage. Engineers were assigned to the Deere & Company Technical Center in February 1970 to develop a multipurpose harvester. A single power unit would power a machine that could replace self-propelled combines and forage harvesters. A unit was built and tested that fall with both alternatives, using a 6600 combine engine, drives, front axle and cab. The project was dropped because there was too much compromise in the forage harvester design, and cost savings to the farmer were insufficient.

Here is a 3300 combine with grain platform operating in grain sorghum. Kansas, Texas and Nebraska grow most of this crop. The 3300 combine, like its 40 predecessor, had limited sales and was on the market less than a decade.

The 4400 combine with a 40 series corn head was popular in the Corn Belt and the East in the early '70s. It was an economical way to harvest four rows of corn, either wide or narrow spaced.

This 6600 combine has a pickup reel operating in a nicely standing rice crop. Most rice combines use spike-tooth cylinders for more positive feeding and better threshing of the tough rice straw. The 6600 was Deere's best selling combine in the 1960-90 period, with 5,451 units sold in 1973.

The SideHill 6600 combine reduced cleaning losses when operating on side slopes. Paddle conveyors replaced the feeder house chain in this design to accommodate the tilted header. It was introduced in 1975, and the following year 654 units were sold.

This Turbo 7700 combine, built in the late '70s, has a rotating air-intake screen that is kept clean by a vacuum arm. The pickup reel on this 218 platform is effective in picking up lodged grain.

The Titan combines, introduced in 1979, included the 6620, 7720 and 8820. The all-new 8820 had six walkers, a 65.5-inch-wide cylinder, a 200-hp engine and a 222-bushel grain tank. Grain platforms as wide as 30 feet were available to utilize this new capacity. Hydrostatic drive was standard on the 8820 in series with a 4-speed transmission. The 6620 and 7720 provided more productivity, too. They had larger grain tanks, and engines with larger displacement and more power. All three models had cabs with a heater. An air conditioner was optional.

The updated 4420 came in 1980 with a larger displacement engine. Power and grain tank capacity were maintained to keep the cost down, as this was now John Deere's smallest combine. This did not save the 4420 from the fate of its predecessors in the '60s and '70s, and it was replaced in 1986 by the Zweibrücken-built 4425. There still is a market for that size combine in Germany and other European countries.

The appetite for more combine capacity in the U.S. and Canada is hard to satisfy. Sales of the 7720 combine passed those of the 6620 in 1980 and were double their amount in 1984. The Titan II line was introduced in 1985. Engine power was increased on each model, with the 8820 reaching 225 hp. The cleaning shoe had up to 15% more capacity from its greater length, faster shake and higher air flow. A feeder-house reverser was standard on the self-propelled models.

Inventors have been intrigued for decades with the idea that rice and wheat could be stripped from the standing straw without having to cut it. This could result in a lighter, simpler combine with more capacity. Major research projects on strippers have been conducted at Davis, California; the International Rice Research Institute in the Philippines; and Silsoe, England. The potential seemed promising, and John Deere engineers worked on strippers in the '70s. Like others, they found that it was easier to strip grain than to save it. Strippers, which work well in standing grain, have difficulty in down grain catching enough of the grain to be acceptable.

Here's a 4420 combine in oats. Although it had more capacity than the original 55, it was too small to satisfy many North American customers in the '80s.

This 7720 Titan II combine is well equipped to meet the challenges of muddy rice fields. It has dual drive wheels and hydrostatic rear-wheel drive to improve flotation and traction.

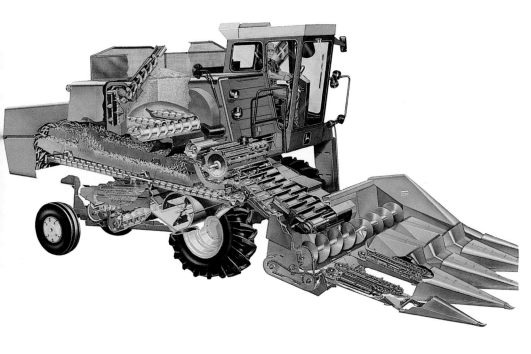

This cutaway of a 6600 combine shows the long augers used to convey grain to the shoe from the cylinder, beater and straw walkers. A fountain auger is used to fill the grain tank from the clean grain elevator.

One of the best combinations for harvesting soybeans in the '80s was the 6620 combine with 653A Row-Crop head, shown here. The 6620 sold well to the corn-soybean-wheat farmer who switched heads several times a year, or even daily.

Here's a 6622 hillside combine operating on a slope near its maximum of 45%. Hillside combines are sold primarily in Washington, Oregon and Idaho. Sales were best in the '70s, with the 6602 having sales of 259 units in 1976.

In 1989, the Zweibrücken 4425 combine was replaced by the updated 4435. This was an ideal combine for the farmer who needed the capacity to harvest four rows of corn or soybeans. Although it may seem small for the '90s, it has a larger engine and grain tank than the huge 105 combine introduced in 1961.

The most important farm equipment introduction by John Deere in the late '80s was the Maximizer line of combines. The 9400, 9500 and 9600 combines were first sold in 1989. Their immediate acceptance by farmers and custom operators is a tribute to John Deere engineering and manufacturing and the local John Deere dealer.

Most customers liked the centered cab and the engine behind the grain tank, as in the 55 combine of four decades earlier. The location of the functional units, such as the cylinder, concave, straw walkers, cleaning fan and shoe, has a heritage that goes back at least eight decades to the stationary thresher.

What, then, has improved in this period? The most obvious is the capacity of the machine. The 9600 easily harvests more than 25 bushels of corn per minute. The capacity of the Maximizer combines is 15 to 20% greater than the models they replaced.

Grain quality has also improved with the new combines. The larger 26-inch-diameter cylinder runs slower over a longer concave to reduce crackage. The most important functional innovation on the combine is the Quadra-Flo cleaning system. Its precleaner helps produce clean grain with minimum losses.

The cab and controls produced the most favorable comments by new owners. The cab is quiet, clean, spacious and comfortable. Visibility is good, and the controls make it easy to become a confident operator. A passenger seat is a popular option.

There is about one self-propelled combine for each seven tractors in the U.S. The eight states of Illinois, Iowa, Minnesota, Kansas, Indiana, Ohio, Nebraska and Missouri have more than half of the self-propelled combines in the U.S. Saskatchewan and Alberta, Canada, each have about the same number of combines as Missouri.

The 9400 combine shares many of the same parts with the two larger models. It is the same width as the 9500 but has shorter (157-inch) straw walkers. It also has a smaller (182-bushel) grain tank. Its 359-cu.-in. engine produces 167 hp.

The 9500 is the favorite size of Maximizer combine, accounting for half of all sales. It is shown with the 843 corn head. Customers like the easy-opening side shields, which provide good access to service points. Maintenance is simplified by one-third fewer chains and belts.

Here's a 4435 combine, with 443 corn head, unloading grain into a parked truck. It has a 24-inch-diameter by 41-inch-long cylinder. The grain tank holds 125 bushels and the engine has 117 hp.

Wheat farmers find the 9600 well matched to their needs. The optional 253-hp engine features a 12-hp power boost when unloading on the go, to compensate for the power used in unloading. Customers like the buddy seat for training new operators and showing neighbors their new combine.

Platforms

Platform selection for small PTO combines was simple, as there were few options. The main decision with the 45 and 55 combines in the early '50s was the width of the platform.

By the '60s, there were more choices. Draper platforms were available for rice combines. On these, the cutterbar was moved forward and a short draper conveyed the cut rice back to the cross auger. The pickup reel had steel fingers to pick up and feed the often lodged crop. Farmers with windrowed grain could add a belt pickup to the conventional cutting platform with the reel removed. The edible-bean grower used a different pickup on a conventional platform that was less likely to pick up rocks.

In the early '60s, platforms were removed from the combine at the pivot point immediately ahead of the cylinder. This was logical then because the early corn heads harvested only two rows. The ears from these two rows were conveyed into the cylinder by twin augers between the rows. With the coming of 4-row and larger corn heads, there was a need for a cross auger similar to grain platforms. Thus in 1964 the feeder house was designed to handle all crops and became an integral part of the combine. This reduced total costs to the farmer who needed both a corn head and a platform.

The 200 series platforms of the late '70s and early '80s were a distinct improvement in design. Attaching the header and its drives was simplified. An enclosed knife drive handled rigid platforms as wide as 30 feet. Flexible-cutterbar platforms reduced soybean losses by conforming to the contour of the ground. Automatic header height control assured close cutting with minimum operator attention.

The 900 series platforms were introduced in the mid '80s. Rigid and flexible platforms were available in widths of 13 to 30 feet. Both used stainless-steel feed plates behind the sickle to improve crop flow to the cross auger.

The 24-foot 224 platform was the widest rigid unit available until the 8820 combine was introduced in 1979. Rigid platform with bat reels are preferred if only standing small grain is harvested. In the latter half of the '80s, rigid platform sales wer less than half of flexible platform sales. The 30-foot 930 was the best selling rigid platform for the Maximizer combines.

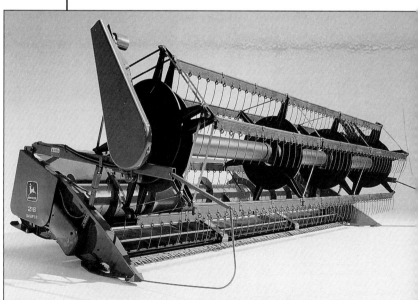

Down, tangled rice is fed positively with this 218 draper platform. The three rubber conveying belts behind the cutterbar feed the crop evenly to the 24-inch diameter cross auger, used on all 200 series platforms.

This 215 platform with finger pickup reel was a good choice for harvesting six narrow rows of soybeans. The pickup reel also helped in picking up lodged grain crops. The hydrostatic reel drive allowed easy matching to ground speeds.

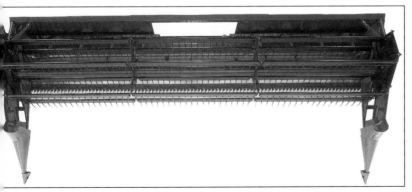

Nylon fingers are used on pickup reels on 900 series flexible platforms. The nylon is durable and does not damage the knife if it is accidentally encountered in rough terrain. Flexible platforms are the most versatile platforms, and thus are sold on more than two-thirds of John Deere self-propelled combines. The best selling size is 20 feet wide, a match for eight narrow rows of soybeans.

The 912 and 914 pickup platforms are designed specifically for windrowed grain. The double windrow, shown, fully utilizes the width of the cylinder and the capacity of the combine. The belt pickup gets under the windrows with minimum shattering of grain. Hydrostatic drive on the pickup permits matching its speed to ground travel.

Row-Crop Heads

Most important innovations introduced by John Deere have been the result of considerable perspiration and some inspiration by engineers assigned to specific crops and machines. The Row-Crop head is the result of diligent effort, but was intended for an entirely different crop. Thus, it could be considered in the same class as vulcanized rubber, penicillin and other commercial successes based on fortunate accidents.

Engineers at the Des Moines Works were looking for a better way to strip cotton. Knowing that in short-fiber cotton they could cut the plant at its single stripping, they developed an experimental unit using rubber gathering belts from Ottumwa Works' forage harvesters. They were impressed with the smooth, positive conveying of the cotton stalks, but were unable to solve the complete stripping process. Rather than considering the whole project to have been in vain, they invited Wayne Slavens, manager of product engineering at the Harvester Works, and Everett Lee, product planning, Deere & Company, to see it demonstrated. The unit was run in a field of soybeans. Slavens and Lee recognized it had the potential to cut the soybean plant low and not shake off the beans.

The Harvester Works followed through with the design, testing, development and production of the 50 series Row-Crop head, and introduced it in 1975. The versatile flexible platforms were being adopted because they saved about 1.5 bushels of soybeans per acre more than rigid platforms. However, Row-Crop heads also were rapidly adopted by farmers with large acreages of soybeans, because an additional 1.5 bushels per acre could be saved. Sales of Row-Crop heads exceeded 3,000 units in 1978 and in 1979.

The John Deere exclusive Row-Crop head is the ideal way to harvest soybeans, sorghum and sunflower. More than half of U.S. soybeans are produced in the states of Iowa, Illinois, Minnesota, Indiana and Missouri. Sorghum is concentrated in Kansas, Texas and Nebraska. North Dakota produces the majority of sunflower in the U.S., with some production also in Minnesota and South Dakota.

Here's a 454 Row-Crop head on a pull-type 7701 combine. The North Dakota spring wheat grower might prefer a pull-type combine for windrowed grain. The Row-Crop head could be used for sunflower or soybeans in that state.

The sleek, long gathering points have a low 20-degree slope to gently lift down soybeans into the gathering belts. The low rotary knives cut the crop with minimum vibration. The cross auger has the same 24-inch diameter as the auger used on platforms.

The 653 Row-Crop head for 30-inch-row soybeans has consistently been the favorite size. Annual sales exceeded 1,000 units in 1977-79.

This 1253 Row-Crop head was offered in the late '80s but had limited acceptance. Recent sales show that farmers wanting more than 8-row capacity prefer the 925 or 930 flexible platforms.

Header attachment was simple and fast, with the introduction of the 200 series platforms. Shown here is a 454 Row-Crop head, a 220 flexible platform, a 215 rigid platform, a pickup platform, and a 643 corn head. Although corn heads and Row-Crop heads appear similar at a distance, corn heads can be recognized by their smaller cross-auger tube and higher auger flights.

Corn Heads

The product development department of Deere & Company was asked to develop a method to harvest corn with less loss. Corn stalks vary widely in diameter, toughness, brittleness, ear height, etc. Corn picker snapping rolls were made aggressive to avoid plugging with stalks and trash. Unfortunately, they also caused considerable shelling as they rotated momentarily against the butt of the ear. Various alternatives were examined. Pickers designed for the sweet corn canning industry were positive in pulling the stalk through but did not let the ear contact the rolls. This principle was used in developing an adjustable snapping bar to keep the ears away from the aggressive fluted snapping rolls.

In the early '50s the University of Illinois was experimenting with combining corn, using a modified cutting platform. Deere ran many stationary tests with a 55 combine to determine if a rasp-bar cylinder would effectively shell corn with all the shucks in place. The result of this combined work was the introduction of the 10 corn head by the Des Moines Works in 1955.

Much was learned about corn head design in the first 15 years before the 40 series was introduced in 1970. The 10, 205, and 210 corn heads fed the ears into the cylinder directly from row-unit augers. The 410 corn head, shown on page 289, was a complex unit with many drives. The 34 series corn head, introduced in 1964, eliminated the complicated drives with its rigid frame and a cross auger that fed into the feeder house chain. It worked well, although in certain conditions the enclosed cross auger plugged with trash, and cleanout was difficult. Another improvement on this series was the use of snapping plates, with ears conveyed to the rear by the gathering chains rather than a separate auger.

In 1967, the 313 and 812 corn heads eliminated the cover over the cross auger. The classic 40 series came out in 1970. In appearance the 40 series differs little from its immediate predecessors. There is no single feature on it that is spectacular, but it is one of the best examples of overall good engineering to be found in any John Deere product.

The 3-row 334 corn head was well matched to the 55 combine and to corn planted and cultivated with 6-row equipment. The ear shields that kept ears from rolling back out the front of the unit were much lower than on the 210 corn head. The development of hybrid corn that stood better and had more uniform ear height contributed to better harvesting also.

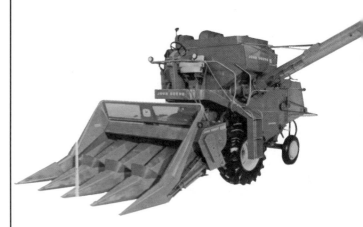

Here's the 435-N corn head for four narrow rows. Less than one-third of 435 corn head sales were for narrow rows in the late '60s.

The 210 corn head was primarily responsible for the demise of corn pickers. Annual sales exceeded 4,000 units in 1963 and 1964. Corn heads cut field shelling losses in half in many instances. Combining corn could start earlier in the season—as soon as the moisture content declined to 30%, if the farmer had a way to dry the shelled corn.

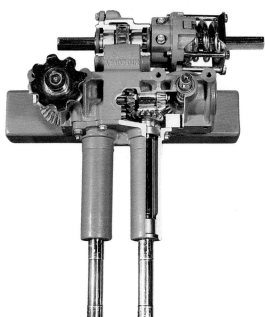

The 812 corn head was designed for the promised yield increases of 20-inch-row corn. The yield increases were hard to detect, but the added costs of the planter and corn head were not. Less than 300 units of 20-inch-row corn heads were sold in the decade following their introduction in 1968. Rear tractor tire widths made it difficult to cultivate 20-inch rows.

The row-unit gearcase is the heart of the 40-series-corn head design. It is much simpler and more durable than its predecessors of the '60s. It is mounted on a toolbar and can be set for 28-, 30-, 36-, 38-, or 40-inch row spacing, depending on the model.

Probably the best evidence of the superiority of John Deere corn heads is the number of them that are seen on non-green combines. A small firm in Nebraska made adapter bundles to fit 40 series corn heads on competitive combines.

Annual sales of John Deere corn heads exceeded 10,000 units in 1973-76. More than 5,000 units of the 444 corn head were sold in 1973. It retained its best seller status through 1978. Since then, the 643 has been the favorite, accounting for about half of all corn head sales. In the '80s, corn head sales were about half of self-propelled combine sales.

More than half of the corn heads on U.S. farms are in the five states of Iowa, Illinois, Minnesota, Indiana and Ohio. There are half as many corn heads as self-propelled combines in the U.S.

JOHN DEERE EQUIPMENT SALES U.S. AND CANADA IN 1975

Farm tractors	50,522
Corn pickers	405
Pull-type windrowers	392
Self-propelled windrowers	3,380
Pull-type combines	713
Self-propelled combines	12,976
Platforms	13,883
Row-Crop heads	706
Corn heads	10,843
Beet harvesters	206
Cotton pickers	688
Cotton strippers	1,136

The 244 corn head is a direct descendant of the original 10 corn head, but is a much superior unit. The low-slope gathering points help salvage corn that is lodged from storms or insect damage. The shield that keeps ears from rolling out the front of the unit is much lower and less likely to interrupt the flow of tangled stalks.

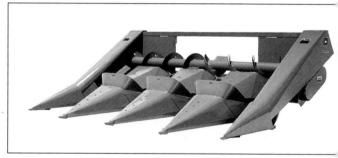

There have been more 444 corn heads sold than any other model. It could be adjusted for 36-, 38- or 40-inch rows.

This 843 corn head shows the more complex outer gatherers used in the early '70s. Ears shaken off the stalks are blocked from going off the front by metal shields on each gatherer and by rubber flaps next to the rows.

The 643 corn head, was the most popular model in the '80s, and is shown on a SideHill 6620 combine. In 1978, for the first time, there were more total rows of narrow-row corn heads sold than wide-row corn heads.

The 1243 corn head harvests 30 feet of corn—12 rows at a time. With typical yields, the 9600 combine fills its 240-bushel grain tank in less than 10 minutes.

Other Crop Harvesters

In the '60s considerable interest developed in storing high-moisture shelled corn in oxygen-free silos. Cattle can also digest corn cobs, so John Deere furnished optional equipment for combines to produce whole- or cracked-kernel corn-cob mix. That option is no longer available, but a corn head can be mounted on a self-propelled forage harvester to get similar results.

Edible-bean versions of the 40 through 105 combines were available in the '60s. They used spike-tooth cylinders and concaves for good feeding and complete threshing of the tangled windrows of beans. A slow-speed drive reduced cylinder speed to 274-400 rpm to minimize cracking.

Castor beans were probably the most fascinating crop harvested in the '60s. They were supposed to become popular and provide additional farm income if only there was a good harvester. Deere developed a unique header to get the bean capsules off the plants. A 55 combine was modified by using a rubber-covered cylinder and concave to rub the three beans out of the capsules. A bucket elevator was used to reduce damage to the beans. Less than 180 units were sold in seven years.

The cash value of peanut production in the U.S. is similar to that of grain sorghum and higher than rice or barley value. John Deere made an entirely different combine for peanuts. Annual sales of the 111 peanut combine averaged less than 60 units in 1965-69. Georgia, Alabama, Texas and North Carolina produce four-fifths of the U.S. crop.

Potatoes are important in the U.S. diet and to farmers as a cash crop. Deere made potato diggers for many years, with the 2-row 30 potato digger having sales of more than 400 units in 1960 and 1961. However, sales faded as farmers switched to potato harvesters to save labor. John Deere had an active project to develop a potato harvester in the late '50s, but decided against producing it, partly because harvesting conditions are so diverse. Key states include Idaho, Washington, California, Maine and Florida.

Whole-kernel corn-cob mix, shown, was obtained by covering many of the openings in the concave and setting the concave clearance too small to let whole cobs through. Cracked-kernel corn-cob mix was obtained by the above means and by increasing the cylinder speed.

Here's the 2-row castor bean attachment, which mounted on a modified 55 combine. Flailing arms struck near the base of the woody castor bean stalks to shake the capsules off. The unit was surrounded by hardware cloth to confine the flying capsules. The stalks passed between stiff nylon brushes that kept the capsules from falling on the ground. Augers on either side of the row collected the capsules.

THE JOHN DEERE *45* EDIBLE BEAN COMBINE

GIVES YOU MORE AND CLEANER BEANS WITHOUT CRACKING

The 45 and other edible-bean combines had special equipment to reduce problems with stones and dirt. For stones, they had a quick-stop for the platform and feeder house. There was also a stone trap at the lower front of the cylinder. Dirt in the beans was reduced by using perforated housings for the clean-grain and tailings elevators.

John Deere was the first to offer self-propelled convenience with the 111 peanut combine. It used two spring-tine cylinders in series to remove all the pods without shelling. The hydraulic-dump tank held 70 bushels. Since peanuts grow just under the surface of the ground, they are dug with a digger-shaker that is similar to a potato digger. After drying in the windrow, they are combined.

The 2-row 30 potato digger was a rather simple machine. The front blade dug under the potatoes. Star wheels vibrated the potato chain to shake most of the sandy dirt through before depositing the potatoes on top of the ground. A cross-conveyor conversion for the digger made it into a stone picker.

Cotton Pickers

Cotton was the last major field crop in the U.S. to be mechanized. There were at least two reasons for this delay. First was the difficulty of selectively harvesting the mature cotton without damaging the remainder. The higher yielding varieties are traditionally harvested multiple times. The second reason is that cotton farmers normally had low-cost hired labor do the picking. Conversely, farm families traditionally furnished at least half the labor used to harvest most of the other major crops.

Wheat was well mechanized in the days of horses and required less than 10 man-hours of labor for an acre's production in the late '30s. Corn had to wait for the tractor picker, so did not reach the level of less than 10 man-hours per acre until the late '50s. Cotton reached that level in the late '70s. Labor used to produce an acre of cotton dropped from an average of 83 man-hours in the late '40s to 8 man-hours in the late '70s.

John Deere entered the market in 1950 with the 2-row No. 8 cotton picker, made by the Des Moines Works. With a well-proven picker unit design, the 1-row tractor-mounted No. 1 cotton picker was added in 1955. Farmers switched to cotton pickers rather rapidly, since picking costs were cut to one-half or less. Thus, the peak sales for Deere pickers in the 1960-90 period came in 1960 when 2,461 cotton pickers were sold. The all-time best seller was the 2-row self-propelled 99 cotton picker with annual sales exceeding 1,000 in 1960-62. It was available as a high-drum picker (20 spindles high) or a low-drum (14 spindles per bar). There were 16 bars in the front drum and 12 bars in the rear drum.

The 1-row tractor-mounted 22 picker was almost as popular, with annual sales greater than 1,000 in 1960-62. It used the same picking drum as the 99 and had similar features. It mounted on 2010, 3010 or 4010 tractors in less time than its predecessors because rework of the tractor was minimized. The tractor was driven in reverse.

The tractor-mounted 22 cotton picker was a good choice for the first-time buyer. It cost less than half as much as self-propelled units because it only had one picker unit, and it used the farmer's tractor for the engine, drivetrain and tires.

The tractor-mounted 12 cotton picker was an attempt to make an economy picker for the small-acreage farmer of the Southeast. Picker bars were 10 spindles high. This lighter unit could be mounted on tractors as small as the 40. Sales never reached 100 units in any of the six years it was sold.

The 699 cotton picker introduced the Jet-Air-Trol system in 1970. The cotton no longer went through the fan, reducing damage to the seed and avoiding embedding trash in the lint. The new chassis had as options a cab and automatic height-sensing of the row units.

The self-propelled 299 cotton picker is shown in the first picking. Picker units had many minor changes to improve reliability. The chassis was lowered to permit the use of a larger basket.

The self-propelled 99 cotton picker was the choice for large farms and replacement buyers. It had wide tires and 34-inch crop clearance. The hydraulic-dump basket held 1600 pounds of seed cotton or 2100 pounds with an extension.

The market for 1-row tractor-mounted cotton pickers dwindled in the latter half of the '60s. Farmers wanted 2-row harvesting capacity. They did not want to tie up one of their tractors, nor take the time to mount the picker.

The 9900 cotton pickers, introduced in the mid '70s, were completely new. The picker units functioned better and were much lighter. The front drum, which previously had 16 bars, now had 12 bars, similar to the rear drum. Aluminum bars were used instead of steel, and the spindles were shorter. Smaller diameter doffers also reduced weight. The cam that controls the spindle angle relative to the drum was redesigned to operate more smoothly. The increased drum speed permitted by this design allowed an increase in travel speed for higher productivity.

The reduced weight of the picker units permitted the introduction of the 4-row 9940 cotton picker in 1981. This was followed in 1990 with the first narrow-row picker, the 9960.

The functional principles of cotton pickers have remained the same since introduction in 1950. High-yielding, long-fiber, open-boll varieties of cotton are defoliated before the first picking. Tapered, barbed spindles enter the plants perpendicular to the row, winding the exposed lint on the barbs. As the cam-positioned spindles pass slowly by the faster-rotating rubber-faced doffers, the cotton lint is removed. Air then transports it to the basket. Before the spindles reenter the row, moistener pads wet them to remove gum and make them grab the lint better.

Annual sales of cotton pickers averaged more than 1,000 units in the '70s. There has been a gradual shift to more high-drum units. In the early '70s, about two-thirds of pickers sold had the 20-spindle-high drums. In the late '80s, this had increased to three-fourths of 2-row cotton pickers having high drums. Low drums (14 spindle) are not offered on the 9960.

Three-fourths of the cotton in this country is grown in the five states of Texas, California, Mississippi, Arizona and Louisiana.

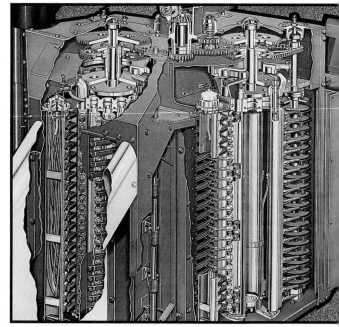

Here's a cutaway of the picker unit used on 9900 series cotton pickers. There are 480 spindles in each row unit. The spindles on the front (left) drum are shown ready to enter the row. The individual bevel-gear drives to the spindles are shown exposed in the picker bar. To the right, spindles are shown passing by the doffers that remove the cotton lint.

The 9910 cotton picker, introduced in 1978, featured refinements on the 9900. The lighter, better designed row units permitted picking speeds of 3.2 mph in first gear.

The 4-row 9950 picker is shown on a road between the near field that is ready for harvest and the far field that is still immature. The wide-stance rear axle provides stability for heavily loaded baskets of cotton.

The 2-row 9930 picker had as standard equipment a 608-cu.-ft. basket with compactors. Options include a 14-inch extension on the basket, and a vane unloading system to split the load for topping off trailers or module builders. Sound-Gard styled cab with heater is standard.

This 9960 cotton picker is equipped to pick five 30-inch rows. The new picker units have 18-spindle bars, with both drums picking from the right side of the row. The picker is shown dumping into a module builder. Seed-cotton storage on the farm has reduced labor peaks in cotton harvesting, because the cotton can be hauled to the gin over a longer period.

Cotton Strippers

John Deere entered the cotton stripper market with a horse-drawn unit in 1930, two full decades before developing cotton pickers. Strippers are much simpler machines because they take the boll as well as the lint, in a once-over operation. In many respects, the tractor-mounted 77 cotton stripper of 1960 was similar to a mounted corn picker. Paired inclined steel rolls removed the bolls from the stalks. However, the rolls did not grip the stalk but rather swept up the stalk, removing the bolls. The bolls fell into an auger, were conveyed to the rear cross auger, and then up the wagon elevator. Green bolls could be saved or dropped back on the ground.

Stripper varieties of cotton have been preferred in the High Plains of Oklahoma and Texas because their closed bolls prevent loss of the mature crop in storms. Stripper cotton has short fibers, while picker cotton has long fibers. Preparation for harvesting also differs for stripper cotton. Harvesting starts after the plant is dead from frost or from application of a desiccant. This eliminates the leaves; it also dries the bolls and weakens their connection to the plant.

When farmers were given a choice of steel or brush rolls in the late '60s, they switched rapidly to the brush design of the 282 stripper. The next choice they made was to have the basket mounted over the tractor—with the 283 stripper, sold in the '70s. The 482 and 484 strippers of the '70s brought self-propelled convenience and 4-row capacity. The self-propelled 7445 cotton stripper, sold from 1986 through 1990, could harvest six 30-inch rows or various spacings of four or five rows. A saw-type cotton cleaner is available for the 7445.

Sales of John Deere strippers were similar to sales of pickers in the '60s and '70s. The demand for the longer staple cotton harvested by pickers caused stripper sales to decline to only one-fifth of picker sales in the late '80s. Annual sales exceeding 1,000 units of the 77 strippers were made during four years in the '60s, and of the 283 stripper during three years in the '70s.

The 77 cotton stripper was the favorite of the '60s. It mounted on 2010 through 4020 and earlier tricycle tractors . Cotton is conveyed to the large wagon by a combination of a belt elevator and a fan.

The 484 self-propelled cotton stripper brought high productivity by harvesting four rows at speeds up to 7 mph. Row units could be adjusted for a variety of row spacings in both solid planting and skip-row cotton. Stripper rolls had rubber flaps alternating with nylon brushes.

The 283 mounted cotton stripper brought comfort and convenience to the operator. It is shown on a tractor with Sound-Gard body and adjustable front axle. It featured brush rolls and had automatic height sensing of the row units.

The 7445 stripper could harvest six 30- or 32-inch rows. With the optional saw-type cleaner, up to 500 pounds of trash was removed per bale.

Harvesting, Overseas

Lanz Combines

The Heinrich Lanz company began building hand-powered stationary threshers in Germany in 1867. During the 1870s Lanz developed a mobile thresher that was driven, and later transported, by a steam traction engine. In 1929 an all-steel thresher, very similar in appearance to U.S. machines of that time, was introduced.

Lanz opened a second factory in Zweibrücken in 1931 to build binders and other harvesting machines. The first prototype combine was built there in 1938.

With the interruption of World War II, it was 1953 before the MD180 PTO combine with 1.8-m (6-foot) cut went into production. The following year Lanz introduced their first self-propelled combine, the MD240S, with a 55-hp Opel-Kapitan engine and 2.4-m (8-foot) cutting width. Both types were initially equipped with sacking platforms, and were painted cream color with red wheels and lettering. Lanz had been building stationary balers and presses since 1927. Most of both types of combines had a low-density straw press attached at the rear. Straw was a very valuable commodity in Europe for livestock.

In 1958 all machines followed the tractor lead and adopted John Deere green, with yellow wheels and lettering. Among the many different models and sizes introduced, the most popular combines were the MD18S, built from 1958 to 1963, and the MD25S, introduced in 1960 with either a grain tank or a sacker. The MD25S remained in production until the introduction in 1965 of John Deere's own design, the new 30 series combines.

During 1961 a design team from the Harvester Works in East Moline, led by Homer Witzel, visited Europe to study the competition and produce a new combine design suitable for Europe. Separation of grain from straw was more difficult than in the U.S. because the straw was longer and remained damp during harvest, due to the weather. The team found that the two largest competitors had taken the John Deere 55 combine design and widened the separator to handle European straw conditions.

An early Lanz PTO combine of the '50s is fitted with a sacker, a straw press and a chaff collector. It is powered by a Lanz Bulldog tractor.

The green and yellow John Deere-Lanz MD25S combine is shown in a typical European scene of the early '60s. The 8.5-foot-cut sacker unit had a straw press (baler) built in. The low-density bales are rather loose and irregular shaped.

A preproduction 730 combine with 105 decals is shown during the 1964 harvest in Norfolk, England.

The smallest of the 30 series combines, a 330 with 8.5-foot cut, harvests a standing crop of wheat.

This 730 combine is shown with a 4-row 435N corn head. Zweibrücken combines were equally at home in small grain, corn, rice and other crops. Note the ladder folded for transport.

The 630 combine was the most popular size in the 30 series. It is shown with crawler tracks, harvesting rice in the mud.

Unique Features on 900 Series Combines

When the 30 series combines appeared, they were a widened and updated form of the classic 55 combine design, which has been followed in all the German-built series to date. In 1965, the first combines announced were the 330, 430, 530 and 630. They were followed in 1967 by the top-of-the-line 730 and the 360 PTO-driven version of the 330. The 430, 630 and 730 were exported to Canada as a lower priced option to the heavier built U.S. combines.

The 30 series remained in production until the major Saarbrücken announcement of September 1972, when the 430 through 730 were replaced by the 940 through 970. The 900 series combines introduced two new concepts, a revolutionary cross-shaker and an exclusive optional Revermatic transmission on the 960 and 970 models. This allowed the operator a conventional gear change by pushing the clutch pedal halfway down, but it reversed the combine's direction if pushed the full distance—a great advantage at corners or in tough crop conditions where plugging was possible. Other new features were Quik-Tatch headers, only 10 daily service points, and an optional hydraulic fore-and-aft reel adjustment.

The smaller 930 combine was added the following year with a 68-hp 4-cylinder diesel engine made by John Deere in Saran, France. For the 1975 season, its 935 replacement kept that engine, while a smaller 925 model was added with a 52-hp 3-cylinder diesel. The remainder of the line changed in 1976 with new Power Pack hydraulic system and larger grain and fuel tanks. Separation improved with the extra (fourth) walker step. Larger chaffer and sieve areas improved cleaning. The 975 featured a Posi-Torq ground drive.

Platforms from 8.5 to 18 feet and corn heads for three to six rows were available. For the 1977 season, the SideHill 965H combine joined the line. The following year the economical 952 combine was added, with four walkers but no cross-shaker. In the fall of 1978, Zweibrücken announced their largest combine to date, the

This 940 combine is equipped with a 343 corn head and grain tank extension. France and Italy are the main corn producing countries in Western Europe. The 940 was the smallest combine using the cross-shaker system.

The SideHill 965H combine saves grain on moderate slopes because the separator remains level. Zweibrücken-designed combines are made for the Chinese market in Jiamusi, China, under a license from Deere & Company.

Key features are shown in this 965 combine cutaway. The cross-shaker is shown below the rotary screen for the engine.

This 955 combine has the top sacker option. A rotary grader was also available on 955 and 955R combines.

The largest combine offered in Europe in 1979 was the 985; it had six straw walkers and a 61-inch-wide separator.

This Zweibrücken-built 1075A combine is shown in Argentina with an 18-foot 218 flexible platform. Combines exported to Argentina and Mexico did not have the cross-shaker.

6-walker 985, making a total of nine models.

In the fall of 1981, the line was again upgraded with the 1052, 1055, 1065, 1068H, 1075, 1075 Hydro-4, 1085 and 1085 Hydro-4 combines. The 125-hp 1072 joined the line for conditions not requiring the power of the 150-hp 1075. The 10 series combines used 10- to 20-foot 200 series grain platforms and 4- to 6-row low-profile 40 series corn heads. The SG2 Sound-Gard cabs made in Bruchsal, Germany, were available.

A Novel Australian Combine

Australia introduced the special 1051 PTO combine in 1983. It had a Chamberlain-designed closed platform and the European 1065 separator. The dry climate produces a sparse stand of short wheat yielding 5 to 10 bushels per acre. Long tapered guards form a closely spaced comb. Losses are reduced if the combine travels about 10 mph so the wheat falls on the platform instead of the ground. The toothed front auger sweeps the wheat into the rear cross auger.

The final major change in Europe, the 1100 series, began in 1987 with the introduction of the 1188. The Hydro-4 version of the 1188 had a 200-hp engine. The line had many improvements in visibility and control for the operator. The curved windshield sloped to the rear and the unloading auger swung forward 110 degrees. Within the cab were a sampling drawer for clean grain and a tailings door.

The 800 series grain platforms were introduced at the same time, with attachments for rape and sunflower. Soybean growers could choose the 900 flexible platform. The lengthened feeder house had a reverser to remove plugs quickly. The Slope-Master cleaning shoe reduced losses on sidehills.

With additions in 1989, the 13-model lineup included the 1133, 1144, 1155, 1157, 1158, 1166, 1166 Hydro-4, 1169H, 1174, 1177, 1177 Hydro-4, 1188 and 1188 Hydro-4.

The 1100 SII series was introduced in 1990 in the three larger models, with more powerful engines and improved optional straw choppers.

The 1051 PTO combine is shown with a 25-foot Australian closed platform. A similar 22-foot version was available as well as a conventional 25-foot 200 series rigid platform with bat reel. Note the 150th-anniversary decals.

The 1177 and 1188 Hydro-4 SII combines had more engine power and green grain tank extensions instead of the previous black extensions.

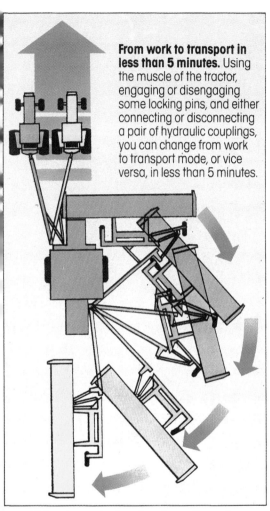

From work to transport in less than 5 minutes. Using the muscle of the tractor, engaging or disengaging some locking pins, and either connecting or disconnecting a pair of hydraulic couplings, you can change from work to transport mode, or vice versa, in less than 5 minutes.

This schematic drawing shows the unique and speedy method of converting the 1051 combine to its transport position.

The Zweibrücken-built 1133 combine is shown with SG2 cab and straw chopper. All 1100 series combines could have 11 rods removed from the chaffer extension to save grain on slopes. The 1133 had a Saran-built 65-hp 4-cylinder diesel engine.

An 1177 Hydro-4 combine is shown with corn head and optional 6000-liter (170-bushel) grain tank.

An 1166 combine and 814 platform, with new straw chopper, harvesting partially lodged grain in Germany. The sicklebars of 800 series platforms could be extended easily for peas, beans, and tall or down crops.

Farmstead

Feed Equipment

Farmstead equipment also has been called materials handling equipment and livestock equipment. Most of it is at least partially related to livestock. Feed equipment is primarily used in livestock production.

The mechanization of field crops was easier and happened earlier than livestock mechanization. The entire 1960 to 1990 period witnessed steady gains in labor productivity in livestock production. Labor hours per unit of production were halved each decade for milk, pork and broilers. More confinement housing was used as livestock production became more concentrated. Labor productivity for livestock production in the late '80s was six times that of the late '50s.

The 200 bale elevator was designed specifically for the one-man haymaking system that stored small bales without hand stacking. It was available in sections to use either as an elevator or as a conveyor in the barn mow. Low in cost, it also was rugged enough to handle full-size bales. In feed equipment it was the best seller, with annual sales exceeding 2,000 units in 1961-65.

The 350 portable elevator was primarily used for ear corn but also could be used for grain or bales. Annual sales were more than 1,000 units in 1961-65.

The 43 power sheller and the 10-A hammer mill each had sales exceeding 1,000 units in 1960. But sales were in a steep decline as corn heads came in and farmers tried to eliminate shoveling.

The 400 grinder-mixer saved considerable labor, relative to its hammer mill ancestors. Annual sales exceeded 2,000 units in 1967-70, but it also was destined for extinction. Grinder-mixers were dropped in the '80s because farmers preferred the quality and convenience of commercially prepared feed.

The 200 bale elevator and conveyor completed the one-man haying system by elevating the small bales into the mow and then distributing them with a movable unloader. Transport trucks were available for moving the elevator.

The 350 portable elevator certainly made ear corn storage easier than shoveling from the tailgate of a flare box. Note the barge box with hydraulic dump. Elevators had optional drives for tractor PTOs, electric motors or gasoline engines.

The 400 grinder-mixer took much of the work out of the daily to weekly chore of grinding and delivering feed to hogs or cattle. A long auger allowed feed delivery without entering the feedlot. Concentrates could be added and mixed with farm-grown grain to obtain the desired ration.

Shoveling ear corn into a 43 sheller was one way to keep warm when the snow was on the ground. Through the '50s and into the '60s, most farms included livestock. This operation may look like hard work, but it is considerably more productive than the hand sheller used a generation earlier.

The 10A hammer mill brought the convenience of portability with its transport wheels and PTO drive. The operator is shoveling ear corn out of a crib. The cyclone collector reduces dust before the feed is dropped into the flare-box wagon, making feed more palatable for livestock.

319

Rotary Cutters

About the only attention pastures received up to the '30s was fencing to keep livestock in or out. In the '30s, the agricultural extension service promoted liming and better varieties of grasses and legumes. Carrying capacity per acre was further increased in the '40s by the use of fertilizer and by mowing to control weeds.

Sicklebar mowers were the only choice for pasture mowing then. Swath boards failed frequently in this rough service. Cattle and sheep kept most brush from starting, except thorn bushes. Locust trees were a problem, and osage orange growth was even more prevalent in fields with hedge fences. Breaking knife sections on thorn bushes was so frequent that some farm boys learned to rivet in new sections in the field.

Highway departments were having the same problem in maintaining right-of-ways. They started using rotary cutters in the late '40s to reduce downtime. Some farmers bought cutters to clear pastures that had been taken over by brush. Those that had reasonably clean pastures found it difficult to justify an additional machine when they already owned a sicklebar mower for their hay crop.

Rotary cutters cut by impact, with a typical blade tip speed of 150 mph. (Home walk-behind mowers have tip speeds of about 200 mph.) As rotary cutters found more uses, various duty levels evolved, with the heavy-duty cutters able to cut 4-inch-diameter trees and brush. Centrifugal force on the swinging blades of even the economy cutters is sufficient to cut 1-inch brush.

The swinging blades offer some protection to the cutter when stumps or boulders are encountered. John Deere's economy cutters have oval-type blade holders to jump stumps. On the rest of the cutter line, bar-type blade holders are built strong enough to handle stumps and large rocks.

Here's a 5-foot-cut pull-type 127 Gyramor rotary cutter operated by a 1010 tractor in mixed hay. It was soon recognized that rotary cutters lost too many leaves from quality hay. A manual crank adjusted height for cutting and transport.

This 5.5-foot cut integral 207 Gyramor rotary cutter is shown in a dense growth of medium size brush. Note the deep deck to provide room for the cut brush. This width was well matched for another popular use, cutting two rows of cornstalks or cotton stalks.

The 13-foot 8-inch pull-type 707 Gyramor rotary cutter, shown in cotton, had two overlapping and timed rotors. It was used primarily for cutting four rows of stalks and secondarily for pastures.

The 205 Gyramor rotary cutter was introduced in 1966 and became John Deere's best-selling rotary cutter in the '70s. This was a light-duty 5-foot-cut unit for grass, weeds and cornstalks.

This integral 7-foot 709 rotary cutter is clipping a pasture for weed control. It also is designed to cut up to 4-inch-diameter brush. This cutter features the all-welded unibody construction for long life. The formed frame is located on top to allow free circulation of cut material under the deck.

The force necessary to cut brush with free-swinging blades is also able to throw objects great distances. Chain shields are needed on rotary cutters when mowing near people or buildings.

Sicklebar mowers cut by slicing and shearing. They must be kept sharp to work, but require little tractor power. The impact cutting of rotary cutters, even when the blades are sharp, requires more than twice the power per foot of cut of sicklebar mowers. Impact cutting soon dulls the tips of the blades if rocks or dirt are encountered. Dull blades will continue to cut, but require added power.

John Deere entered the rotary cutter market in the '50s with the 127 and 207. They were both well accepted, with sales exceeding 5,000 units of the 127 in 1960 and in 1961. Introduced in 1961, the 6-foot 307 utility cutter was quite popular in the '60s. The 7-foot 407 Gyramor cutter, introduced in 1962, established Deere in the heavy-duty market with its ability to clear 4-inch-diameter brush. The 2-rotor 707 Gyramor cutter followed in 1964, with stalk-cutting capacity increased to four wide-spaced rows.

An entirely new line of cutters was started in 1972 with the introduction of the heavy-duty 09 series (509, 609, 709). The 5-, 6- and 7-foot cutters built on the reputation of the 407, with many improvements. Blade attachment, for example, was improved by using larger keyed bolts to prevent loosening. Annual sales of the 709 cutter exceeded 1,500 units in the '70s. The 15-foot pull-type 1508 flex-wing cutter followed in 1975. This high-capacity cutter's ability to cut 2-inch brush has increased its sales over the years for clipping pastures and cutting stalks.

The regular-duty 506 and 606 cutters followed in the late '70s. The 506 was the best seller in the early '80s, with about one-third of all cutter sales. Economy cutters were introduced in 1986, with their sales exceeding all other cutters since then.

Over the years, the preference in single-rotor cutters has shifted almost completely from pull-type units to integral units. Rotary cutters are sold throughout the country, with some added concentration in the humid Southeast.

Options make the 2-rotor 10-foot 1008 cutter one of the most versatile cutters sold. Integral and center-pull units provide a wider cut than single-rotor cutters for clipping pastures or cutting stalks. The offset model is a favorite in orchards and groves. Options include limb deflectors, tree guard bumper wheels, and hooked blades to mulch prunings.

The 3-rotor 14-foot 1418 cutter, shown in cotton, is the answer for chopping four wide rows or six narrow rows of stalks. Integral cutters can be used on row-crop tractors. Pull-type units make rattle-free turns up to 80 degrees by use of a constant-velocity front universal joint.

The 3-rotor flexible 15-foot-cut 1508 cutter provided one-sixth of all John Deere farm rotary cutter sales in the late '80s. Its flexibility is an advantage for cutting roadsides and rough-terrain pastures. It is equally well liked for cutting corn, cotton or grain sorghum stalks.

The regular-duty 506 rotary cutter was a good choice for farmers with no pastures to clip. It could keep weeds under control around the farmstead and in grass waterways. Farmers now keep their roadsides cut more frequently, for better appearance.

The economy 503 cutter is the right choice for the homeowner with a small acreage. The one-piece welded body shields the rear and improves mulching. A shear bolt protects the gearbox from damage. The 403, 503 and 603 cutters each outsell the 709, the best selling cutter for farm use.

Farmstead

Rear Blades

The 6-foot-wide 80 rear blade was introduced by the John Deere Spreader Works in East Moline, Illinois, in the '50s. The Spreader Works was the source of most farmstead equipment in the early '60s, but it was closed in the late '60s. Rear blades and most other farmstead equipment are now made at the Welland Works in Ontario, Canada.

Rear blades are handy, versatile tools found on most farms. Uses vary with tractor size and include farmstead maintenance, feedlot scraping, snow removal, and ditch digging and maintenance. They are used to maintain roads within the farm and to touch up other gravel roads between gradings.

As more sizes of tractors were introduced, both lighter and heavier rear blades were added to the line. All five recent John Deere blades can be operated in forward or reverse and adjusted for angle, tilt, offset and pitch.

The original 80 blade and the following 80A were the best sellers, with annual sales averaging more than 2,200 units for more than a decade. In the '70s, annual sales of the 45, 78, 88 and 115 rear blades each averaged more than 1,000 units. Sales of the 45 have continued strong and amounted to one-third of total rear blade sales in the late '80s. Sales of rear blades peaked in 1979 at 7,429 units.

Rear blades are unrelated to any specific crop, so they are spread throughout the country. Their use is higher on livestock farms.

It became practical to do some light earth-moving jobs with farm tractors as their power and weight increased. Degelman Industries was one of the pioneer manufacturers of agricultural bulldozers. John Deere purchased a range of dozers from them from 1971 through the first half of the '80s. The best seller was the 534, with sales greater than 1,000 units in 1975 and 1976. Dozers could do better work than rear blades on most jobs, but were more expensive and more difficult to put on the tractor, so were not as popular.

Here's an 80 rear blade grading a gravel country road. Its 6- to 9-foot blades made it popular with all utility tractors.

The 45 rear blade has been the favorite for more than a decade. Its clean, simple design is matched to tractors up to 55 hp. John Deere blades of all duty levels have high reliability.

324

The medium-size 534 agricultural bulldozer, shown on a 4440 tractor, was the best selling model. Its rugged frame transferred forces back to the tractor drawbar.

The 78 rear blade was popular throughout its long life, from 1965 through 1982. It was available in 6-, 7- and 8-foot widths. Like other Deere blades, it could be reversed, angled, tilted, offset and pitched.

This 115 rear blade is the choice for use with 100-hp tractors. It is shown scraping a concrete cattle feedlot. Its optional hydraulic offset and tilt make it well equipped for ditching.

Farm Loaders

Loader convenience and productivity have come a long way since John Deere's first model in 1940. It was a cable-operated unit, with the tractor parked like a backhoe. The bucket faced away from the tractor, as in steamshovels of that time. The front-mounted 25 loader, introduced in 1946, showed that the engineers had recognized some of the basic features needed in a loader. It was still cable operated but got its live power from the flywheel to avoid excess clutching and shifting. It had a mechanical bucket trip, the industry standard of the day. And it had parallel lift, a deluxe feature even in 1990.

Introduced in 1955, the 45 loader established John Deere as a major loader manufacturer. That reputation has grown over the years, with Deere recognized for their loader design for the past two decades. The 45 was a simple, rugged, straightforward design. It used a mechanical bucket trip but had live hydraulics for the boom. It had 3,000 pounds of breakaway force and could lift 2,750 pounds 10 feet. It was matched to all row-crop tricycle tractors, and was one of the first Quik-Tatch loaders on the market. Annual sales exceeded 4,000 units in 1960-62.

The 46 loader, introduced in 1962, set the basic layout for most current loaders with its hydraulic bucket control and its placement of masts in front of the operator's platform. This design permitted easier access for the operator and was easier to fit on wide-front-end tractors.

It was the later 48 loader that targeted the mass market. Sales exceeded 6,000 units each year in 1969-71. Part of the high sales can be attributed to its fitting 2010 through 4020 tractors. Some tractor durability problems led to better matching of loaders to tractors after that. Introduced in the first half of the '70s, the 146, 148 and 158 loaders remain in the line. Since then, about 50,000 units of the 148 loader were sold. It fit 2520 through 4455 tractors. Total John Deere loader sales peaked at 16,670 units in 1979.

This 145 loader is shown moving a farm gate, another muscle job of livestock farming. This model could maneuver in tight spaces, because the bucket was close to the tractor. The swept-back front axle shown also helped make short turns.

Cattle farmers like the versatility of the 146 loader. It is shown with a manure fork and grapple picking up a round bale in the field. The 146 fits 4-cylinder utility tractors from the 2020 through the 2755.

The 45 loader was designed for John Deere's 2-cylinder tractors, but got additional productivity with 3010 and 4010 tractors. These New Generation tractors had faster hydraulic lift and their Syncro-Range transmissions allowed much faster and easier maneuvering.

The 46 loader set many of the basic design features of the next 25 years. The masts were located forward for easy access to the operator's platform. Bucket dump was hydraulic. The boom was arched to fit over adjustable front axles. It could be operated with the remote controls on the dash or with its own valves.

Here's a very economical and productive team for handling fruit or vegetables. The 48 loader with 27 forklift is placing a pallet of apples on a truck. A 27 forklift on the 3-point hitch doubles the carrying capacity of the tractor and counterbalances the loader.

Farm loaders were originally called manure loaders because that was the dirty, heavy-lifting job they were primarily designed to do. They had a mechanical-trip manure fork with few options. Farmers found additional uses, so materials buckets and forklifts were added to the line. Materials buckets were better for handling soupy manure. In 1990, tines added to materials buckets were more popular than manure forks.

Handling round bales is now a major chore for loaders. The most positive way is with the optional round bale/silage grapple attached to a materials bucket with tines. The round bale fork, optional on 148 through 280 loaders, is a fast, economical way to handle round bales.

There are many other loader uses. The 59 fork-lift easily handles pallets. The telescoping boom attachment to forklifts extends the reach of the loader. Non-farmers have also found the loader a much-used tool. The 75 loader, designed for Yanmar-built 850 through 1050 tractors, had good sales through the '80s.

The big loader news in the '80s was the intro-duction of the John Deere 200 series in 1981. These deluxe loaders featured mechanical self-leveling, and the Quik-Tatch system permitted attaching and detaching without tools. Quik-Tatch buckets also made bucket changes faster and easier. The loaders had the strength and clearance to be used on tractors with mechanical front-wheel drive. Single-lever control is standard on the 245 and optional on the 265 and 280. Lift height varies from 11 feet 6 inches on the 245 to 14 feet 10 inches on the 280. Lift capacity is 2,530 pounds on the 245 and 4,580 on the 280. The 245 has a breakout capacity of 3,855 pounds and the 280 has 6,220 pounds.

Sales of the 245 loader surpassed those of the 148 in the mid '80s. Since then, the 200 series loaders have provided more than one-third of Deere farm loader unit sales. Loaders are found on most farms in the U.S. and Canada. Their concentration is highest in the Midwest, where most of the spreaders are located.

This 75 loader on a 950 tractor was as handy for the small-acreage homeowner as the larger loaders were on farms. Tapered pins were used in the boom and bucket pivots for easy attachment and removal from the tractor. Note the bucket-level indicator on the left boom. John Deere loader buckets also have a flat top to help the tractor operator determine bucket pitch.

This 158 loader is poised to grab a big load with its silage grapple and materials bucket with tines. The 158 is matched to tractors from the 4000 to the 4630 without mechanical front-wheel drive.

Here's the popular 148 loader on a 4230 tractor. The rear blade adds ballast to the rear of the tractor and is ready for scraping the concrete clean. The 148 is often placed on an older tractor and left on.

This 265 loader is making good use of its lift height and reach in stacking round bales three high. The round bale/silage grapple also can be used to stack bales on end in sheds where there is no need to shed rain and snow.

The Quik-Tatch mounting of a 245 loader on a 2750 tractor with mechanical front-wheel drive is shown. Changing buckets is even simpler.

Spreaders

Horse-drawn spreaders were developed more than a hundred years ago, but some farmers were still pitching it off wagons 50 years ago, one forkful at a time. Many early spreaders had wooden cross slats mounted on chains, so the whole floor conveyed the load to the rear. A beater at the rear shredded the manure so it would not drop in large clumps.

The first spreader designed and made by John Deere, the Model "A," was introduced seven decades ago. It featured the unique "beater on the axle." Anyone who had ever loaded a spreader by hand could see the benefits of the low loading height. It also retained the large-diameter wheels for easy rolling in soft ground.

John Deere entered the '60s with three spreaders having annual sales exceeding 2,000 units. They were the Models "L," "R" and "N." Each had a main beater, an upper beater and a widespread. Total spreader sales dropped from 13,416 units in 1960 to 6,530 units in 1967 and have been even less since. The tradition of most farms having some livestock declined rapidly in the '60s.

Dairy barn cleaners became more popular in the '60s and needed a different spreader design. In 1961 the 33 dairy spreader was introduced, with a low, wide box and a single combination beater-widespread. It and its successor, the 34, were immediate hits and had annual sales exceeding 3,000 units in 1963-69. The 40, 54, 660 and 680 spreaders that followed used the same general design with a single beater-widespread unit and a conveyor chain.

Conveyor chains have always been a trouble spot on spreaders. Daily hauling by dairy farmers accelerates rusting, and winter hauling results in added loads from freeze-down. The original malleable chain had given way to steel detachable chain by the '60s. The 54 spreader used a log chain. Later spreaders offered T-bar chains as an option. The 350 and 550 spreaders, sold by Deere in 1990, were made by Badger Northland. They used a T-rod conveyor chain on a fiberglass/plastic-coated floor.

More than 4,000 of the Model "R" spreader were sold in 1960 and 1961 because it offered high value to most farmers. The 95-bushel ground-driven unit provided five conveyor speeds, controlled from the tractor.

The 40 spreader was in the line for almost two decades because it matched the needs of many dairy and other livestock farmers. It provided 175-bushel capacity and a low loading height, less than 4 feet.

The 76-bushel ground-driven Model "L" spreader was the right choice in 1960 for the farmer with limited livestock. It had a narrow, straight tongue that permitted 90-degree turns for maneuvering in tight quarters.

The 134-bushel PTO-driven Model "N" offered new productivity to the livestock farmer. The arch over the top beater was too high to pass under many dairy barn cleaners. Many spreaders were sold without tires so they could be fitted with 20-inch used truck tires.

Here's the 118-bushel 33 dairy spreader. Note the low profile and the even spreading by the single combination beater-widespread. An optional rear pan is shown.

John Deere eliminated the troublesome conveyor chains with the 450 Hydra-Push spreader introduced in 1980. Two Minnesota dairy farmers originated the idea of pushing the load off with a movable panel operating over a plastic-covered floor and sides. Manual cleanup after daily spreading is practically eliminated because the fiberglass-reinforced plastic surfaces remain relatively clean.

Dairy farmers quickly recognized the merits of this design, and the 450 became the best selling spreader in its second year. At the end of the '80s, the larger 780 Hydra-Push spreader was the best selling model. Hydra-Push units provided more than half of Deere spreader sales in the last half of the '80s, and the trend continues. They were originally designed and built by the Welland Works, but in recent years these spreaders have been built by MacDon Industries in Winnipeg, Manitoba.

More than half of the manure spreaders on U.S. farms are in eight states: Iowa, Wisconsin, Minnesota, Illinois, Ohio, Missouri, Indiana and Nebraska.

The 870 flail spreader was built for John Deere by Farmhand. Flail spreaders are simple, and handle soupy manure and frozen manure well. The earlier 70 flail spreader had sales of 527 units in 1969.

JOHN DEERE EQUIPMENT SALES
U.S. AND CANADA IN 1975

Farm tractors	50,522
Grinder-mixers	1,042
Rotary cutters	6,992
Rear blades	3,812
Agricultural bulldozers	2,144
Farm loaders	11,758
Spreaders	3,290
Wagons	11,013
Posthole diggers	1,068
Skid-steer loaders	815
Implement carriers	1,240

The big-capacity 54 spreader was introduced in 1970 and had annual sales exceeding 1,000 units for a decade. It was low enough to back under most barn cleaners and large enough to meet the needs of beef farmers. Spreading rate was rope-controlled from the tractor.

Here's the spreader that handles the wide variety of manure produced on dairy farms. The 450 Hydra-Push spreader is shown making the daily haul from the barn cleaner. The optional hydraulic endgate helps confine sloppy manure so it is not lost on the way to the field. The movable panel is shown about one-third of the way to the rear.

This 400-bushel 680 conveyor spreader was the largest unit John Deere ever sold. Tractor drawbars must be strong for spreader operation, because the weight balance shifts as the load moves to the rear.

This 780 Hydra-Push spreader is a favorite with feedlots (shown) and dairies. Unloading is faster with this design, because the operator has a wider range of unloading speeds and cleanout is more positive.

Wagons

Wagons predate most farm tools other than plows. They were used in the cities to haul freight and on the farm to haul hay, grain, cotton, lumber and livestock. Many farmers in the '20s and '30s had a low steel-wheel wagon for hauling hay and a high wooden-wheel wagon for hauling grain.

Switching wagons to tractor power was a simple matter of shortening the tongue and adding a clevis hitch. As with most other implements, tractor power prompted a major redesign in wagon gear. Rubber tires, frequently old car tires, permitted traveling at tractor speeds. Automotive steering replaced fifth-wheel steering for better stability. Wagon gear changed from wood to steel for the heavier loads and higher speeds with tractor operation.

John Deere entered the '60s with a proven line of four wagon gear models, known for their straight tracking and light draft. The best seller was the 4-ton 953. The heavier, wider-stance 1064 was recommended for use with forage boxes and other heavy loads. All models had telescoping reaches that permitted adjusting wheelbases from about 7 feet to 11 feet. The tie-rod was located back of the front axle to protect it from being bent when obstructions were hit. Optional rocking bolsters were located at the rear for better load stability during turns.

The wagon line was redesigned in the early '70s. A new C-section reach gave a flexible yet tight connection between the front and rear axles. Heavier tie-rods were moved to the front of the axle. Tire sizes and carrying capacities increased. The new 1275 had six wheels, with the rear ones mounted on walking beams. Capacity was 12 tons at field speeds.

John Deere wagon sales declined slowly in the '60s and '70s from 20,408 units in 1960. Annual sales of the 7-ton 1065A exceeded 5,000 units in 1972-79. Sales declined more in the '80s as round bales and module builders replaced the handling of hay and cotton by wagons. In recent years, about one-third of the wagons sold by industry are used with gravity boxes and one-fourth with forage boxes.

Here's the 7-ton 1065 wagon sold in the '60s. Its adjustable-width stakes are shown without bolsters. Factory-installed brakes were available, either using the truck braking system (shown) or surge tongues.

The wagon line was updated in 1988 with capacity increases of 10 to 40 percent. The 11-ton 740 wagon is shown. Other models include the 7-ton 700, 9-ton 720 and 14-ton 770.

This 5-ton 963 wagon shows two options available in the '60s. Spring bolsters cushion heavy loads on roads. Four-wheel surge brakes help in stopping the wagon behind a pickup truck. John Deere wagons have a reputation for straight trailing on highways.

The six-wheel 1275 brought 12-ton capacity to the John Deere wagon line. The walking beams distribute the load equally to the four rear wheels and provide a smoother ride. This is the gear for heavy forage boxes.

This 10-ton 1075 wagon shows the improvements made in the wagon line in the early '70s. The C-section reach provides a strong, tight connection between the front and rear axles. Heavy tie-rods are located in front of the axle. The telescopic tongue made it much simpler for one-man hitching. There is an optional fixed bolster in front and a rocking bolster in the rear.

Outside Manufactured Products (OMP)

Sales volumes of some farm equipment items do not justify the expense of designing and manufacturing by large companies. Yet dealers may need these products to offer a more complete line for certain crops, operations or systems.

John Deere used some outside manufacturers for equipment throughout the 1900s, and chose this as a strategy in the '70s. A separate organization called OMP was established, with most of the functions of a factory other than manufacturing. The customer and dealer received product support similar to that offered on products made by John Deere, including operator's manuals and parts catalogs. All products were tested by Deere engineers, and changes were made where needed for safety, function, durability or appearance. A few products were designed by Deere and made by others, such as the recent Hydra-Push spreaders.

Some OMP items were successful in the market, others were not. The OMP function was transferred to the related factories in the '80s. Some factories continue to supplement their own manufacturing with some products made outside. JDM (John Deere Merchandise) remains very active. It provides John Deere dealers with a wide variety of shelf items such as oil, batteries, hand tools, etc.

The 33 bale elevator was obtained from Portable Elevator Mfg. Co. following the closing of the Spreader Works. Sales exceeded 1,000 units in 1972 and 1973. The 31 posthole digger, made by Arps Mfg. Co., has had sales greater than 1,000 units in 15 of the years it has been sold. It is sold by agricultural and consumer products dealers.

Skid-steer loaders were first sourced from Owatonna Mfg. Co. Inc., and more recently from New Holland. Sales reached 1,249 units in 1979, with sales through agricultural, industrial and consumer products dealers. The 201 implement carrier, built by The Donahue Corp., had annual sales exceeding 1,200 units in 1975-79. The 500 grain cart, sold in the late '80s, was made by United Farm Tools, Inc.

The 33 bale elevator worked equally well for full-size bales, as shown, or for smaller bales thrown into high-sided wagons. It could be used to elevate bales into barns or for stacking outside in drier climates.

The nimble maneuverability of the 575 skid-steer loader is valuable in the older buildings of hog or dairy operations.

The 31 posthole digger takes the back-breaking work out of fence making. Augers are available in 6- to 24-inch sizes. The digger can be removed and the remainder used as a crane with 1250-pound capacity.

The 500-bushel 500 grain cart is the way to keep ahead of the Maximizer combines. It holds two dumps from the 9600. It can be used to haul directly to farm storage or to a roadside truck to haul greater distances.

The 8x28-foot 201 implement carrier is used to transport planters and other wide implements on roads and through narrow gates. The bed of the carrier rests on the ground for easy loading of the implement. The bed is then hitched to the tractor and slid back up the frame and over the wheels, ready for transport.

Consumer Products

Walk-Behind Equipment

Most of John Deere's consumer products have been lawn and grounds care equipment since the company entered that business in 1963 with introduction of the 110 lawn and garden tractor. Recreation equipment was a venture, with snowmobiles sold from 1971 through 1983 and bicycles sold from 1974 through 1976. The Horicon Works in Wisconsin has been devoted totally to consumer products since 1969.

For homeowners, the most regular and frequent outdoor chore is cutting grass. Traditionally, it was done by pushing a reel-type mower that cut a 16-inch swath and worked best when cutting at about a 1-inch height. This is too low for a healthy Kentucky bluegrass lawn in the Midwest. In the '30s, weeds were prevalent in lawns because the grass was cut too low, fertilizer was not used and chemical control had not begun.

Institutions started using rotary mowers in the '40s, and their use spread to homeowners in the '50s. Some homeowners were convinced that reel-type mowers were the only way to have a good lawn. They soon observed that their neighbors with rotary mowers no longer had buckhorn, and that higher mowing resulted in less brown grass during dry weather. Improved results, combined with faster mowing from a wider cut, caused a rapid switch to rotary mowers.

Most of these mowers had the blade mounted directly on the vertical crankshaft of the engine. This was easy with 2-cycle engines, but lubrication and carburetion had to be modified to get 4-cycle engines to run with a vertical crankshaft.

John Deere entered the walk-behind mower market in 1970 with units purchased from Toro. This was followed by making them at Horicon. The next source was Lawn-Boy. Total Deere sales of walk-behind mowers exceeded 20,000 units in five of the years in the '70s. A completely new line of walk-behind mowers, with cast decks and rear grass baggers, were made by John Deere in 1989-90 in their new factory in Greeneville, Tennessee. Sales of these mowers have been considerably better than those of previous models.

Horicon introduced this deep, cast-deck self-propelled 21-inch mower in 1978. The optional rear grass bagger gave better weight balance and caused less interference around shrubs.

This pressed-steel-deck, push-type 20-inch mower, designed and built at Horicon, was sold in the mid '70s. It had a Briggs & Stratton engine and is shown with the optional side-mounted grass bagger.

This self-propelled 21-inch mower is one of six models purchased from Toro in the early '70s. It has the optional side-mounted grass bagger attached.

Here's the cast-aluminum-deck, self-propelled 21-inch 12SB mower, made by John Deere in Greeneville, Tennessee. The 12SB features a 2-cycle 4-hp engine, a blade brake clutch, and a 5-speed transmission. The rear grass bagger is standard and the front thatcher is optional.

The 320 snow thrower, built by Jacobsen, has a 2-cycle engine. Adjustable vanes direct the snow thrown by the 20-inch-wide impeller equipped with two rubber paddles.

Most homeowners in the United States and Canada own a walk-behind mower. Annual industry shipments of these mowers exceeded 5,000,000 units annually during the last half of the '80s, making them distinctly the most popular outdoor power equipment for the homeowner. More than two-thirds of these mowers are push-type, with 20-inch units providing half of push-type shipments. Half of the self-propelled units have a 21-inch cut.

Power has also simplified two other home chores, snow removal and gardening. Three walk-behind snow blowers were introduced in 1970. The number of snow throwers and blowers shipped is more variable than that of other equipment because it depends on seasonal snowfall. Annual industry shipments of snow throwers and blowers in the late '80s were in the range of 300,000 to 600,000 units. Almost two-thirds of these were snow throwers, which have a single-stage rotor that not only picks up the snow but also throws it. John Deere's most popular unit of this type was the 3-hp 320 snow thrower, with sales of 15,906 units in 1979.

More than three-fourths of industry shipments of snow blowers in the late '80s were for units less than 26 inches wide. However, John Deere's best seller was the 26-inch 8-hp 826 snow blower, with 7,541 units sold in 1979. Snow blowers have two stages. An auger picks up the snow and a blower discharges it. Snow blowers have more power, work in deeper snow, and move the snow farther than snow throwers.

Walk-behind tillers have taken much of the backbreaking work out of gardening. They can be used for once-over tillage in the spring, and later for cultivation. Annual industry shipments of walk-behind tillers averaged about 300,000 units in the last half of the '80s. Slightly more than half of these are front-tine tillers. Deere's most popular unit was the 6-hp 24-inch 624 tiller, shown on page 85, with sales exceeding 12,000 units in 1974 and 1975. The 4- to 6-hp units are the most popular industry sizes for both front and rear-tine tillers.

Walk-behind tillers and snow throwers and blowers have higher sales than tractor-mounted units because they are more maneuverable and require no hookup.

The TRS27 snow thrower entered the market in 1990. This 8-hp 27-inch-cut thrower is made by Noma Outdoor Products, Inc. Four models of snow throwers have wheels for propulsion and two use the Deere exclusive Trax track-drive system.

Here's the economical 2-hp 16-inch-wide 216 tiller. It can eliminate much of the hard work in a small flower or vegetable garden. Introduced in 1978, it was still in the line in 1990.

The 826 snow blower is an ideal way to handle deep snows like this. Its powerful blower sends the snow far enough to help eliminate snow buildup along driveways or walks. This unit, designed and built at Horicon, was the most popular Deere snow blower in the '80s.

The 3.5-hp 324 tiller preceded the 216 in the '70s. A reverse gear allowed the operator to back out of tight spots without heavy lifting.

The rear-tine 820 tiller was made by Magna American in the mid '80s. Rear-tine tillers are often used by nurseries. They are heavier and more expensive than front-tine tillers.

Riding Mowers

Many customers need more mowing capacity than a 21-inch walk-behind mower offers. Several considerations enter into the selection of mowing equipment, such as amount of landscaping and the size and contour of the lawn. Many people are willing to spend about an hour per mowing. Typical coverage in acres per hour by various mowing equipment is .3 to .4 for walk-behind mowers, .7 to .9 for riding mowers, .9 to 1.1 for lawn tractors, 1.0 to 1.3 for residential front mowers and 1.3 to 2.0 for lawn and garden tractors. If the lawn is irregular or if the grass is bagged, the area covered per hour will be less.

Customer preference makes the final decision. A rear-engine rider provides excellent visibility and is good for maneuvering around shrubs and trees. Recent John Deere riders have a tight 17-inch turning radius. A riding mower takes less storage space and costs less than most lawn tractors with comparable power and cutting width.

John Deere entered the rear-engine riding mower market in 1970 with the 55, 56 and 57. The 6-hp 56 was an immediate success, with sales exceeding 15,000 units in 1972 and 1973. Horsepower increased in the 1975 update, with the 8-hp 68 riding mower having sales of 15,941 units in 1978. These were rather quiet units, as the engine was completely enclosed. Deere offered an even quieter mower in the early '70s, the electric 90 riding mower.

Industry shipments of rear-engine riding mowers varied from 260,000 to 375,000 units in the last half of the '80s. Deere sales in this period were better than in the '70s. The line was updated in 1983 and again in 1987. The RX75 riding mower has been the best seller since then. It featured an electric-start 9-hp engine and a 30-inch cut. All six models had valve-in-head Kawasaki engines and variable-speed V-belt drive. Cutting width varied from the 26-inch single-blade RX63 to the 38-inch double-blade SX95.

This 7-hp 57 riding mower was the top of the line in the first half of the '70s, with a 34-inch cut. The optional front blade, shown, was designed for light snow removal. The transmission had two forward speeds and reverse.

Here's the quiet electric 90 riding mower. Three 12-volt batteries powered the traction motor and the motor on each of the two blades. Each charge provided about an hour's mowing time, or almost enough to cut an acre with the 34-inch mower.

The RX75 riding mower had a variator drive (V-belt) that offered infinite speed adjustment up to 5.4 mph. The riding mower is shown with the optional front thatcher.

The 8-hp 68 riding mower, introduced in 1975, had three important new features. The engine compartment was completely enclosed to reduce noise for the operator and the neighborhood. The 30-inch mower deck was designed to work with a rear grass bagger, shown. The 5-speed shift-on-the-go transmission permitted easy matching of speed with mowing conditions.

A window improved the operator's view of the filling of the grass bagger on the 8-hp S82. This was the best seller in the mid '80s.

Lawn Tractors

Lawn tractors have been improved so much that they now lead industry shipments of riding equipment used for mowing grass. In the 1986-90 period, lawn tractors provided 63% of shipments, rear-engine riding mowers provided 25%, and lawn and garden tractors provided the remaining 12%. Annual shipments of lawn tractors averaged 783,000 during this period, and continue to grow; the other two remain about steady. Most of these lawn tractors have 10 to under 14 hp.

John Deere helped start today's huge market for lawn tractors when they launched the 60 lawn tractor in 1966. The 60 was an intermediate design at that time. It had a more rugged frame and drivetrain than competitive front- or rear-engine riding mowers. Thus, it could handle snow removal with a snow thrower or a front blade. It was simpler and lower in price than lawn and garden tractors because it was not designed for tillage.

Deere sales of the 60, 70 and 100 lawn tractors from 1966 to 1977 were a fraction of the sales of the 110 and 210 lawn and garden tractors. The peak in that period came when the 70 lawn tractor sold 6,603 units in 1970.

Deere lawn tractor sales took a sudden jump after a completely new line was introduced late in 1978. The 11-hp 111 lawn tractor had sales of 12,034 units in 1979 and was the market leader until it was replaced in 1986. These tractors had new mowers designed to work well with rear grass baggers. The 111 used a 38-inch mower and the 8-hp 108 had a single-blade 30-inch mower.

Two important introductions came in 1981. There was a new hydrostatic-drive 111H in addition to the easy-to-shift, 5-speed-transmission 111. The second model was the 16-hp 116. The lawn tractor line now had the convenience and power that previously had been available only in lawn and garden tractors. Deere sales of lawn tractors passed lawn and garden tractor sales in 1983. Sales received another boost in 1989 with the additional STX suburban tractor line.

The 60 lawn tractor provided a new low-cost way to mow grass and remove snow. Safety was a main design consideration. Starting the engine required shifting to neutral, disengaging the mower or snow-thrower drive, and turning a key.

Lawn tractor sales boomed with the new 11-hp 111 lawn tractor. It had an easy-to-shift 5-speed transmission control on the right fender. Appearance rivaled lawn and garden tractors and provided a kinship with John Deere farm tractors.

Here's the 16-hp 116 lawn tractor, which provided the homeowner with a new level of power. It is shown stirring out clippings and dead grass with its thatcher. The hydrostatic 116H was introduced in 1982.

Here's the STX38 suburban tractor, which set new sales records in 1989 and 1990. It has a 12.5-hp Kohler engine with overhead valves. It is shown with the optional rear bagger on the 38-inch mower. A 38-inch snow thrower is also available.

The 160 lawn tractor was introduced in late 1985 and reached new heights in annual sales. Its sleek fiberglass hood enclosed the engine compartment for noise reduction.

Lawn and Garden Tractors

Two-wheel walk-behind garden tractors were available in the '30s, '40s and '50s for the serious vegetable gardener. Plows, cultivators, sicklebar mowers and carts were common implements. When 4-wheel riding lawn and garden tractors appeared in the '50s, they got considerable use in gardening as well as mowing with mid-mount rotary mowers.

John Deere entered the market in 1963 with the 110 lawn and garden tractor. Customers liked the variator drive in series with a 3- or 4-speed transmission. The principle of this variable-speed V-belt drive had been well established in Deere combines. Sales of the 110 tractor reached 22,280 units in 1966. Sales of matching equipment that year, relative to 110 tractor sales, were 92% for the 38 rotary mower, 23% for the 42 front blade, 20% for the 80 dump cart, 14% for the 30 rotary tiller and 12% for the 36 snow thrower. Equipment matching in recent years is more difficult to determine because there are so many models that can be mixed and matched. It appears that relatively more mowers, about the same number of snow throwers, and fewer blades, tillers and dump carts are being sold.

A redesigned line of lawn and garden tractors came out in 1974. The 200 series models retained the proven variator drive and the 300 series had hydrostatic drive. The hydrostatic-drive 400 was introduced as the top of the line in 1975. Annual sales of John Deere lawn and garden tractors exceeded 30,000 units in 1967 and continued at that level through the '70s. The best selling model shifted from the 110 to the 210 to the 17-hp hydrostatic-drive 317 in the late '70s. The 2-cylinder 18-hp 318 was the best selling Deere lawn and garden tractor through most of the '80s.

The 200 series tractor line was all new again in 1988. The top-model 18-hp 285 has a 2-cylinder, vertical-crankshaft, liquid-cooled Kawasaki engine. The 322 tractor has a liquid-cooled gasoline engine. The 332 and 430 tractors have liquid-cooled diesel engines. In 1990, Deere's 10 models of lawn and garden tractors had 14- to 20-hp engines. Institutional and commercial users have replaced homeowners as the main buyers.

The 14-hp 140, introduced in 1967, was John Deere's first hydrostatic-drive tractor, and had three hydraulic circuits, an industry first. It is shown with a hydraulically controlled mid-mounted blade. The earlier 110 tractor is shown on page 82.

The 12-hp 212 tractor had the added power needed to bag grass clippings, making it the most popular model in the 200 series for a decade.

Here's the 17-hp hydrostatic-drive 265 lawn and garden tractor. It was one of four models introduced in 1988 featuring overhead-valve Kawasaki engines. Time required for hookup or removal of the snow thrower, tiller or front blade was reduced to 5 minutes or less.

The 400 tractor brought power steering and individual wheel brakes to the lawn and garden tractor industry. The added power available in 400 series tractors allows them to operate mower decks from 38 to 60 inches wide. John Deere lawn and garden tractors, like their farm tractor relatives, retain higher trade-in values than competitive units.

The popular 318 tractor provided about one-fourth of John Deere sales of lawn and garden tractors in the '80s. It was well matched to the needs of the institutional or commercial user.

THE MILLIONTH LAWN AND GARDEN TRACTOR is being completed by workers at the John Deere factory in Horicon, Wisconsin. It took the company 14 years to reach the half-million milestone, but only 7 years to reach the million mark, since it began building homeowner tractors in 1963. Deere cites a population shift to the rural countryside as one trend that created a demand for more horsepower and wider mowers.

Gear-Drive Compact Tractors

The best selling John Deere lawn and garden tractor size progressed from 7 to 18 hp between 1963 and 1990. These tractors developed a good reputation among commercial and institutional users. But these professionals used tractors more hours per year than homeowners, and frequently asked for "more tractor" in several respects.

For example, a more productive tractor should have a liquid-cooled diesel engine for durability. For loader operation, it should have a longer wheelbase, more weight, larger tires and mechanical front-wheel drive. It should have a husky Category 1 3-point hitch for a variety of implements. The part-time farmer also needed a tractor; good used tractors with 3-point hitches were hard to find at a reasonable price in the '70s.

To meet these demands, John Deere introduced the 22-hp 850 and 27-hp 950 in 1978. They were built by Yanmar in Japan, where this tractor size was very popular on their smaller farms. They were rapidly accepted in the market, with 4,132 sales of the 950 in 1979. The line was extended with the 33-hp 1050 in 1980 and the 14.5-hp 650 and 18-hp 750 in 1981. The 40-hp 1250 was added in 1982 as a farm tractor, followed by the 50-hp 1450 and 60-hp 1650.

The Yanmar-built tractors have provided excellent reliability. But farm usage of the 850 through 1050 was less than expected. Cutting grass was the dominant use for all 650-1050 compact tractors. Almost all 650 and 750 tractors were sold with mid mowers. Less than one fifth of these tractors used loaders, tillers or snow blowers. About three-fourths of the 850 and 950 tractors had mid mowers; about one-third of them had loaders. Mid mower use decreased with the 1050, and loader use increased to about one-half of these tractors. As size increased, use of rotary cutters and 3-point-hitch rotary mowers also increased. Other implements sold for these larger tractors included box scrapers, rear blades and many other farm tools.

Industry sales of tractors under 40 hp in the U.S. held up much better than farm tractor sales in the '80s. They reached 60,464 units in 1987 and were 42,040 in 1990.

Here's the 14.5-hp 650 compact tractor, the smallest in the Yanmar-built line. It has a 17-hp engine, the way lawn and garden tractors are rated, but as a compact tractor it is rated by PTO hp, just like farm tractors. The 650 has been used primarily with the 60-inch mid-mount rotary mower shown. The mechanical front-wheel drive helps traction and steering with turf tires.

This 750 tractor is shown with the 60-inch 261 rear-mounted grooming mower. This mower is readily removed from the 3-point hitch to free the tractor for other uses.

This 870 tractor shows one of the important features of the 70 series introduced in 1989—the ability to have multiple implements on the tractor. It is shown with the 72-inch mid mower and the 660 tiller.

The 970 tractor with 72-inch mower is the right answer for large areas requiring quality mowing. The continuous live PTO permits the operator to stop the tractor without stopping mower rotation.

One of the major uses for the 1070 tractor is with the Quik-Tatch 80 loader. Farm tractor tires and mechanical front-wheel drive provide the traction that helps get a full load each time.

Hydrostatic-Drive Compact Tractors

Customers liked the Yanmar-built gear-drive compact tractors but they also liked the convenience of the hydrostatic drive on the Horicon-built 300 and 400 series lawn and garden tractors. Sales of John Deere hydrostatic-drive lawn and garden tractors passed those of the gear-drive units in 1980.

Horicon Works engineers were then asked to start with a clean sheet to design the ideal hydrostatic-drive compact tractor line, based on their experience with Yanmar and lawn and garden tractors. It was to have plenty of power and be more user friendly for implement attachment and operation. The result was the 655, 755 and 855, introduced in 1986, and the 955 in 1989.

These tractors use liquid-cooled 3-cylinder diesel engines from Yanmar, similar to those proven in the gear-drive line. They use a Sundstrand 2-range hydrostatic transmission. Planetary final drives and wet disk brakes are used, as in John Deere farm tractors. A 2100-rpm mid PTO provides positive power and easy hookup for mid mowers. Hydrostatic power steering is standard.

Mechanical front-wheel drive (MFWD) is standard on the 955 and optional on the others. The MFWD bevel gears at the front wheels permit steering and increase clearance under the axle. In 1986-88, most buyers chose the MFWD option on 655-855 hydrostatic-drive tractors and 650-1050 gear- drive tractors. The adoption rate of MFWD on compact tractors is much higher than on farm tractors. (The 3155, 4755 and 4955 are the only farm tractors with mechanical front-wheel drive sales of more than 50%.)

Hookup of implements to the compact tractors is quick and easy. The four caster gauge wheels on the mower deck make it simple to slide under the tractor. The loader's parking stand functions automatically, resulting in attaching or detaching the loader in a couple of minutes. The mower, loader and tiller can be left on while any one of the three is in use.

This John Deere hydrostatic-drive 655 tractor has a 60-inch mower deck and hydraulic-dump material-collection system. The collector mounts on the 3-point hitch and has its own engine. The hopper's contents can be dumped into trucks or wagons more than 5 feet high.

Here's a 755 tractor preparing a seedbed in a single pass with a 550 tiller on the 3-point hitch. Five front suitcase weights balance the tiller and better utilize the mechanical front-wheel drive. The mid mower has adequate ground clearance to be left in place.

Mechanical front-wheel drive is standard on the 955 compact tractor. Note the slender front differential housing. In addition to the brake pedal on the operator's right, there are two lower pedals for speed and direction control.

The compact 855 is a very maneuverable loader tractor for cleaning out horse barns. Hydrostatic drive is ideal for all loader operation but especially valuable in tight quarters. Regular tires and mechanical front-wheel drive improve mobility for loader operation.

This 755 tractor shows its clean lines and uncluttered platform. Roll-Gard ROPS and seat belt are standard. Turf tires, rear wheel weights and 3-point hitch are optional.

Snowmobiles

Children for centuries have eagerly awaited the first snow to go sledding. Adults joined in this anticipation with the advent of the snowmobile. Some tracked snow vehicles had been used by industry for years but it was the '60s before practical machines were developed for the consumer. The design that evolved had two skis spread in front for steering and a long rubber-faced track behind for propulsion. All controls were on the handlebars so the driver could sit or kneel behind the engine. The well-cushioned seat was long enough for a rider, too.

John Deere entered this fast-growing (more than 100 brands) recreational market in 1971 with the 400 and 500 snowmobiles. They were an immediate success, with 7,352 units of the 400 sold in 1972. Snowmobiles used high-speed 2-cylinder 2-cycle engines to get both high power and light weight. The engine was mid mounted on the 400, between the operator's legs. A variable-speed V-belt drive connected the engine to the tracks. The drive was similar to that used on combines or lawn and garden tractors, except the ratio was adjusted automatically by engine speed and torque. The track had bogie-wheel suspension.

Deere snowmobile sales peaked at 30,879 units in 1974. The top of the Blitz-Black line was the JDX8 with a 440-cc 42-hp engine located at the front. A slide-rail suspension was optional. The nylon slide bar used snow as a lubricant against the steel-lined rubber track.

Air cooling by fan was used on most Deere snowmobile engines. The Spitfire, sold first in 1977, used free air flowing over exposed cooling fins of the engine. The Liquifire snowmobiles, introduced in 1975, used flying snow to cool a long-tank heat exchanger under the seat. In 1976 John Deere won the prestigious 500-mile race from Minneapolis, Minnesota, to Winnipeg, Manitoba, with a liquid-cooled machine.

Several color schemes were used on snowmobiles. A candy-apple green was tried experimentally across the entire consumer products line, but was not adopted.

This JDX8 snowmobile has the optional slide-rail suspension, used on all later models. Shock absorbers also have been added to the skis for a better ride. The air intake for the front-mounted engine is shown.

Here's the 440 Liquifire snowmobile. The liquid-cooled engine, enclosed in the low hood, was quieter than air-cooled engines.

This 400 snowmobile got John Deere off to a running start. Rough terrain was smoothed by the spring-suspended skis, bogie wheels and rear idler.

The 440 Trailfire snowmobile set a sales record of 8,219 units in 1979. Note that the deer in the John Deere logo is leaping to the right (forward).

The Spitfire snowmobile used free-air cooling for its 2-cylinder engine, saving the power normally used by the cooling fan. Note the colorful striping on this lightweight machine.

Front Mowers

John Deere entered the lawn and grounds care business in 1963 as an extension of the farm equipment business. Engineering and manufacturing were similar but the products were smaller. A major difference was the amount of purchased components. Most Deere farm equipment uses engines and transmissions manufactured by Deere. Consumer products rely on purchased engines and transmissions.

Marketing of consumer products was started through the farm equipment organization and dealers. Farmers were, and still are, key customers. But there are many potential customers in the urban and suburban areas. By 1990, about 1,300 consumer products dealers had been added to serve non-farm customers. The total number of John Deere dealers selling consumer products was about 3,000, since most agricultural dealers also sell consumer products. Unlike some competitors, all John Deere consumer products dealers service the products they sell.

John Deere had two decades of reputation for quality consumer products and superior dealer service when the front mower line was introduced in 1984. Acceptance was rapid and the line grew to six models in 1988. The F935 front mower has provided over one-third of all Deere front mower sales. It has a liquid-cooled 3-cylinder Yanmar diesel engine. All six models have power steering, hydrostatic drive and differential lock. The basic design of the 50-, 60- and 72-inch mower decks was available from previous Horicon products.

One of the crucial decisions made during development was the use of rear-wheel steering on the front mowers. Self-propelled combines have used this design for decades. It provides better control on hillsides and in other difficult traction conditions, compared with caster wheels at the rear.

The superior trimming ability of front mowers was extended to the homeowner in 1989 with the F510 and F525 models. In 1990, the F525 outsold the six commercial front mowers, indicating its rapid acceptance by the residential user. It has a novel 17-hp Kawasaki engine and hydrostatic transmission designed as a unit. It uses a 48-inch mower.

The F911 front mower is shown with the optional ROPS and canopy. It is designed for either 50- or 60-inch mowers. A rear weight improves stability and steering. Traction on wet grass can be improved with the optional hydraulic weight-transfer system.

This F930 front mower, with 24-hp air-cooled gasoline engine, is operating in terrain that needs the stability of rear-wheel steering. Uneven cutting height from wheel tracks is reduced by placing the mower in front of the propulsion unit, and using pneumatic gauge wheels.

Here's the popular 22-hp diesel F935 front mower. It provides long-lasting dependability to commercial and institutional users. A 60-inch rear-discharge mower is available, as well as 60- and 72-inch side-discharge models.

The F525 brought the advantages of front mowers to the homeowner. A simple crank, shown in black, adjusts cutting height from 1 to 3.5 inches. The engine-transmission unit is located low for good stability. Options include a grass bagger and a snow thrower.

The 48-inch unit is John Deere's best selling size of walk-behind self-propelled commercial mower. These mowers have higher industry sales than front mowers due to their maneuverability and lower cost.

Other Consumer Products

A variety of other outdoor power equipment rounds out the consumer products line. Some of it is made by John Deere and some of it is purchased. All products benefit from Deere's large network of dealers, known for their after-sale product support.

The AMT 600 all materials transport was introduced in 1987, followed by the AMT 622 in 1989 and the AMT 626 in 1990. These units have sold well to institutions, landscapers and farmers. They are made at the John Deere Welland Works in Ontario, Canada.

Chain saws were sold from 1970 through 1984. The original line was supplied by DESA Industries and the later line by Echo, Inc. The Echo-built 50V sold 12,180 units in 1979.

The Horicon-made 80 dump cart was sold from 1965 through 1990. Annual sales exceeded 6,000 units in 1967-69.

In 1987 John Deere added several products specifically for golf courses. The 22 greens mower, supplied by Tsuchiya in Japan, has sold best. The 1200 bunker and field rake, supplied by Commuter Industries, Inc., has also sold well since its introduction in 1989.

The AMT 626 is a most useful workhorse for landscape nurseries. Features include low loading height, mobility, maneuverability, little soil compaction, and easy operator entrance and exit.

The 50V chain saw was the ideal size for farmers and many other users. It had a 16-inch sprocket-nose guidebar and was powered by a 44-cc engine.

JOHN DEERE EQUIPMENT SALES U.S. AND CANADA IN 1975	
Walk-behind mowers	21,666
Walk-behind snow blowers	7,553
Walk-behind tillers	16,026
Riding mowers	21,294
Lawn tractors	3,642
Lawn and garden tractors	35,661
Sweepers	5,871
Chain saws	14,792
Snowmobiles	20,612
Bicycles	59,008

This 1200 bunker and field rake provides "tender loving care" to a golf course sand trap. Regular grooming of sand traps helps maintain uniform difficulty of the course. This rake is also used to smooth ball-diamond infields.

The precision-built 22 greens mower provides the daily manicuring required by quality golf courses.

Here's the 80 dump cart, a perennial favorite for more than two decades for hauling behind John Deere lawn and garden tractors.

The No. 4 backpack blower replaces the rake for collecting leaves in the fall. Echo, Inc. is the supplier of this 2-cycle-engine blower.

Industrial Equipment

Crawler Dozers

A crawler dozer is probably the most versatile earthmoving machine worldwide. It is used on many jobs such as clearing, leveling, ditching, backfilling, grading, spreading, landscaping and forest fire suppression. Rear-mounted rippers, scarifiers and winches make the dozer even more valuable.

John Deere's first dozer combination, the "MC" tractor and 61 bulldozer, found almost instant acceptance in the forestry industry. Loggers liked the compact power and maneuverability of the "MC" and used it for everything from trail clearing to log skidding and stacking. Different applications led to various blade widths, inside and outside mounts, and fixed or adjustable tilt and angle.

John Deere's most popular innovation was the T-bar-controlled, all-hydraulic dozer. Introduced in the late '50s, the 64 dozer could be raised or lowered, angled right or left, and tilted up or down, all on the go and with one hand. Landscape contractors loved it and road builders found it ideal for working the slopes around overpasses. Contrary to popular belief, crawler dozers and motor graders require the most skilled operators of the crew. Owners and operators appreciate any design that makes the job easier and more productive.

The greatest John Deere contribution to operator ease and crawler productivity is the Dual-Path hydrostatic drive on the 750 and 850 dozers. With an independent hydrostatic pump and motor for each track, power is proportioned during turns, saving time and fuel. Levers or pedals control the speed of each track without steering clutches or brakes. On turns or straightaway, optimum speed is continuously matched to the load for maximum efficiency. This permits the operator to concentrate entirely on moving the dirt.

Annual sales of John Deere crawlers averaged more than 5,000 units in the '60s and '70s, and peaked at more than 7,000 in 1973. The 1990 crawler dozers ranged from the 60-hp 400G to the 165-hp 850B.

Here is an early JD350 with an all-hydraulic blade. Note the T-bar control lever in the operator's right hand. The shorter lever to the right controls the ripper.

The popular JD450-B crawler with outside-mounted blade does a bit of slope dressing. The two short levers to the right control blade tilt and the rear-mounted ripper.

A full blade of dirt is being rolled by this JD850 crawler dozer. This unit has the optional pedal steering. The operator has his right hand on the blade control lever and his left on the direction-reversing control.

This 650G crawler dozer has additional woods protection over the radiator grille, with limb risers and screening around the operator.

This JD750 with outside-mounted blade is clearing some dirt and rocks. The hydrostatic drive precisely matches speed to the work load for best efficiency.

Crawler Loaders

The "MC" crawler probably was the catalyst that led to John Deere's entry into the industrial equipment market. This tractor, with bulldozer or loader, was in the right place at the right time. The "MC" crawler with 61 dozer was an instant success. The crawler loader took a bit longer, since the crawler used a short 3-roller track, lacking in fore-and-aft stability for effective loader operation. Later models provided longer 4- and 5-roller track options, and several allied equipment loaders were approved. These units were well accepted as the principal machine for the small contractor and as a handy cleanup unit for larger operators.

Since forward/reverse direction changes dominate crawler operation, there was an immediate demand for a quick and reliable direction reverser. Over the years, Deere engineers developed thoroughly reliable reversing systems to match the various sizes of crawlers. The ultimate in crawler powertrains is the hydrostatic drive used in tractors over 100 horsepower, previously described for crawler dozers. A single-lever loader control provides raise, lower and float for the boom, and load and dump operation for the bucket.

The crawler loader is an extremely versatile tool, with many attachments and bucket designs. A multipurpose bucket can be used as a dozer, scraper, clam or bottom-dump device. A side-dump bucket permits close-quarter work where space or traffic lanes restrict maneuvering. Rear-mounted rippers, scarifiers, backhoes, cable plows, and many other special attachments add to the value of this machine.

The crawler loader is a rugged performer, working in areas too difficult for wheel units. Welded mainframes serve both the loader and the crawler, permitting rapid removal of modular components. Undercarriages have been improved with through-hardened tracks and pins that are sealed and lubricated.

A pioneer in small crawler loaders, John Deere in 1990 offered four models from 70 to 140 SAE net horsepower.

The JD350-B crawler loader provided a lot of work capacity for its size and price. Wheel tractor owners wished for a crawler, and this machine was a popular choice.

This 450C crawler loader was sized just right for close-quarter cleanup and finish work. Backfilling around green concrete requires a delicate touch, and the 450's high-low-reverse transmission made it easy.

Truck loading is a common use for crawler loaders. This 455G has plenty of reach and lift height to easily dump into the center of this 10-wheeler dump truck.

Here's a JD855 making rock loading look easy. Hydrostatic drive makes turning from the pile to the truck and back again quick and easy. This is a foot-steer loader, so the operator can keep his hands on the loader control and the direction reverser.

This hydrostatic-drive 755B loader makes easy work of this excavation. The fully enclosed operator's station includes a heater, air conditioner, and pressurization to keep out dust and other airborne particles.

Backhoe Loaders

Early tractor production from the Dubuque Works, though intended for the agricultural market, was well suited for industrial applications. The Model "M" and later models had the size and power needed for backhoes. They were welcomed by small owner-operator contractors who were enjoying the postwar surge in building. Backhoe manufacturers were looking for partners, so companies such as Henry, Davis, Pawnee and Pippen became allied suppliers for John Deere.

In the late '50s, John Deere became the first full-line manufacturer to design and build its own backhoes and loaders. A common hydraulic system was used for the backhoe and loader. Market share improved with this no-compromise design, which was balanced in power and stability for a specific tractor. Customers and dealers appreciated the single warranty source for the tractor and equipment.

Deere's first integral backhoe loaders, introduced in 1971, were the JD310, JD410, JD500C and JD510. These machines represented all that was good in backhoe loader design. Many of their components had been field proven thoroughly for years. Backhoes and loaders were designed for specific tractors. The integral design was stronger because the three units reinforced each other. Hydraulic lines were simplified, and they were routed in protected areas. There were no exposed lines in the entire length of the backhoe boom, dipperstick or bucket. Loader lift cylinders were mounted inside the boom arms for protection.

The hydraulic system used a variable-displacement pump that provided cooler operation, fuel economy, and smoother control for work in tight quarters. Two-lever backhoe control, single-lever loader control, and the hydraulic direction reverser were all exclusive features when first introduced. They were still used by John Deere in 1990. The five backhoe models in 1990 had digging depths of 14 to 18 feet and the loaders had lift capacities of 4600 to 7100 pounds.

This early model JD310 backhoe loader is working in fairly tight quarters, but the closed-center hydraulic system provides smooth, easy control.

Note the clean hydraulic hose routing on this 310C backhoe. There are no hoses exposed along the entire length of the boom, dipperstick, or bucket. The loader bucket is rolled forward for an additional anchor for the digging set.

The long reach of the extendable dipperstick allows this JD410 to clean a ditch and load a truck while blocking only one traffic lane. The dipperstick is shown in the retracted position over the truck.

Here's a JD500 equipped with a demountable backhoe working on a bridge-building project. This multipurpose rig has several other uses on the job.

This 710C is John Deere's largest backhoe loader. It has an SAE digging depth of 18 feet and a loader bucket capacity of 1.75 cubic yards. This tractor is equipped with a cab and mechanical front-wheel drive.

Industrial Equipment

Elevating Scrapers

Horse- or mule-drawn scrapers helped open the West by making the fills for the railroads. Slips were pulled by a team, and the Fresnos, which held about one-third yard, used three or four horses. Digging and dumping were controlled manually. In the '30s, construction equipment pioneer R.G. LeTourneau developed a rubber-tired scraper that combined speed and power.

In the early '50s, elevating scrapers appeared on the market. John Deere became allied with Hancock, a scraper manufacturer in Texas. The first 2-axle scrapers and later gooseneck combinations were powered by the largest 2-cylinder diesel farm tractor. Its transmission could not be shifted on the go, and top speed was limited to 12 mph. All controls except the steering wheel required use of the right hand. In 1959, Deere designed and manufactured its own matching pair, the 400 elevating scraper and the 840 tractor.

The real breakthrough took place in 1963 with the introduction of the JD5010 scraper. This machine had power, shift-on-the-go speed, a well-matched scraper bowl, and operator comfort and convenience. Old-line manufacturers, who initially ridiculed the elevating idea, began to work on their own designs. Elevating scrapers require no pusher to load, haul or dump. Elevator flights passing over the cutting edge move the dirt up and back for complete filling of the bowl. This pulverized dirt spills and spreads more evenly when dumped. Elevating scrapers require about 12 to 14 hp per yard of capacity, while conventional scrapers require 50% more.

John Deere's innovative designs continued. Hydrostatic elevator drive allowed operators to match elevator speed to soil and working conditions for fast, complete loading without stalling.

In 1990, the 11-yard 762B and the 16-yard 862B scrapers had a manually controlled 6-speed power-shift transmission. An optional microprocessor control automatically shifted the transmission to match speed and power for loading, hauling and dumping.

Here's a team of 5010 scrapers on a road building job. The 5010 was the first John Deere scraper with a transmission that could be shifted on the go, for good haul speed. Note the PTO driveline for the elevator.

Replacing the 5010, this JD760 appeared in 1965. The stable, 3-axle design was popular on land-leveling jobs. The scraper could be removed and the tractor used as a prime mover for a sheepsfoot roller, water wagon or other applications.

The 762A scraper set new industry standards in the 11-cubic-yard class. Hydrostatic elevator drive, John Deere exclusive automotive-feel position-responsive steering, and power-shift transmission were combined for excellent productivity and operator comfort.

This JD860-A, the leader in the 16-yard class, has a full load. Note the texture of the material spilling over the side. Such dirt can be spread at any thickness desired in the fill area, for easy compaction.

4-Wheel-Drive Loaders

When John Deere entered the 4-wheel-drive loader market in 1968 with the JD544, it again showed a flair for innovation. Conventional 4-wheel-drive loaders used a wheel-steered, straight-frame design and needed a great amount of room to maneuver. The JD544 was one of the first to use articulated-frame steering, turning 40 degrees to the left or right. The tight 16-foot turning radius revolutionized the industry.

Hydraulic design was equally innovative. A single lever controlled the loader boom and bucket with the closed-center hydraulic system. Later loaders featured automatic boom-lift kick-out. The operator simply preset the required lift height and then was relieved of this repetitive chore as he loaded trucks or hoppers. Bucket dig position could also be preset.

Planetary final drives and hydraulic wet-disk brakes are used on all wheel-driven tractors in the John Deere industrial line. This durable field-proven design runs in a bath of oil and keeps out dirt and water. Brakes require no adjustment. Some machines include a differential lock.

Three transmissions are used, matched to various size loaders. Each provides smooth direction reversing and easy speed control. Engine retard assists in providing good downhill control.

John Deere 4-wheel-drive loader sales have grown in this large market of many sizes. The JD544 series have had several years with more than 1,000 loaders sold. In 1990, the Deere line included six sizes from 75 to 216 hp and buckets from 2 to 5 cubic yards. Larger buckets are available for snow and light materials. In addition, there are multipurpose buckets with clamshell action and side-dump buckets. Various forks are available for loading boxcars and for work in automotive salvage yards. A hydraulic coupler is available to permit the easy interchange of a variety of unaltered attachments.

As one of John Deere's smaller loaders, the 444C is still big on production and features. It has articulated steering, inboard planetary drive and brakes, and a variable-speed hydrostatic transmission with 2-speed Hi-Lo power shift.

This 544D is loading out of a stockpile. The articulated steering turns 40 degrees right or left for quick maneuvering. Only the front section with the loader turns, so the operator experiences little side-to-side movement as he rides in air-conditioned comfort.

No matter which member of the Deere loader family you look at, clean design is evident. Hydraulic lines on this 644D loader are hard to see. They are all well concealed and protected, yet readily accessible for service.

Here's a JD646-B compactor in a typical landfill. It does a thorough job of spreading and compacting each load. A compactor is far more than just a loader with compactor wheels and a modified bucket. Considerable additional shielding is necessary to prevent damage from dangerous objects, and additional cooling is required.

This JD844, introduced in 1978, was the largest of the 4-wheel-drive loader line. It was frequently used in gravel pits or quarries with large-capacity trucks such as this belly dump.

Motor Graders

The first road grader appeared about 130 years ago. Levers controlled the blade suspended beneath a horse-drawn farm wagon. In 1879 a young road builder, J.D. Adams, developed wheel lean to oppose the side draft of the material being moved. His company introduced the first self-propelled motor grader in 1928.

Four decades later, in 1967, John Deere introduced the next major change in grader design, the JD570. Its articulated frame steering provided a short (18-foot) turning radius, about half that of conventional designs. Better maneuverability meant inside corners and small areas could be worked. Frame steering permitted the front wheels to be offset from the back to straddle windrows, get better traction or improve stability on slopes.

The other main innovation on the JD570 was the total hydraulic control of the blade. Blade pitch was included as well as one-hand control of right and left blade lift. For the first time, the operator could stay in his seat and in 60 seconds have the blade in either the right or left high-bank cutting position.

The original JD570 brought many new ideas that remain in the line. It had a differential lock and a direct-drive power-shift transmission with eight forward and four reverse speeds.

A later popular and productive feature was hydrostatic front-wheel drive, available on all but the smallest model. It could be manually engaged by the operator or automatically engaged when drive wheels began slipping. On steep slopes, front-wheel assist permits a heavier cut before side-slip occurs. It also helps avoid getting stuck in the muddy spots often present in road work. John Deere graders are especially popular for snow plowing, with their front-wheel assist and frame steering.

Front-mounted scarifiers or rear-mounted rippers for difficult surface conditions are common attachments for graders.

In 1990 John Deere offered seven models of motor graders, from 90 to 185 hp.

Bank cutting and shaping are common uses for graders. This blade is working a slope of about 45 degrees. A vertical slope can be easily worked by retracting the upper and extending the lower lift cylinders, shown here.

John Deere graders with front-wheel assist excel at snow removal. Various snow wings are available, as well as the front blade shown.

Hydrostatic front-wheel drive makes this heavy cut possible while turning. Sidehill stability is improved and work capacity increased significantly with front-wheel assist. The all-hydraulic motor grader pioneered by John Deere remains the industry standard.

This big JD770 grader is moving a full blade of dirt. The operator's right hand is on the right and left blade-lift control levers for frequent adjustment. The left hand is free for steering or other blade adjustments. This unit is equipped with a rear ripper as well as a front scarifier.

Here's a JD670 finish-grading a wide swath. The articulated mode places the drive tandem on an unfinished surface. Any wheel slippage or other damage to the surface will be removed on the next pass. Note the excellent view this operator has of the area ahead.

369

Excavators

John Deere was well established in the backhoe market when they introduced the all-hydraulic JD690 excavator in 1969. Both machines are used to dig holes or ditches. The JD690 increased productivity by digging deeper, reaching farther and working faster than back-hoes. The 360-degree swing of excavators allows operation in tight quarters for loading trucks and placing spoil. They are ideally suited for basement digging. Their high lift capacity, or craneability, allows them to do many other construction jobs.

The JD690 was easy to operate, like its backhoe counterpart. Two levers controlled all digging functions, and two pedals handled propulsion of the tracks. Productivity brought rapid acceptance of the JD690. When only one function was in use, the output of the two hydraulic pumps was automatically combined. The convenient hydraulic swing system on Deere excavators eases the work of the operator and helps him be more productive. When the swing lever is released, rotation comes to a smooth stop, and remains braked in that position.

The engine on John Deere excavators is used to drive hydraulic systems for digging and propulsion. Two-speed track propulsion was introduced on the JD690-B in 1973, with a crawler-type undercarriage. Track-type excavators travel only about 2 mph maximum, so they depend on trucks for transportation to jobsites.

Wheel excavators offer speed and mobility. In the early '80s, the 690 excavator with wheels was sold to the military for airport runway repair. In 1986 the 595 wheel unit was introduced, followed by the 495D in 1989. Their power steering and good brakes make it practical to operate in city traffic at 20 mph.

In 1990, there were three mini excavators, seven larger models and two medium wheel models. Digging depths ranged from 9 to 30 feet. Various options were available for buckets, dippersticks and tracks. Excavators are made at the Davenport Works, in Japan, and by Deere-Hitachi at Kernersville, North Carolina.

The largest excavator in the current line, the 992D-LC, will dig nearly 30 feet deep and reach about 44 feet. It weighs just under 100,000 pounds. Hydraulic controls are pilot operated for low effort and precise metering. Engine speed automatically returns to idle when no power is being used.

This 595 wheel excavator retains all the digging features that made Deere excavators popular, but is quicker in transport than track units. Hydrostatic drive, power steering and power brakes provide an automotive feel when moving from job to job at about 19 mph.

This JD890 is loading a 10-wheel truck with rocks. Long reach and ease of control make truck spotting simple and increase production.

The mini excavator first gained popularity in the European market, and the idea soon spread. Relative to backhoes, the mini offers the advantage of 360-degree swing, and tracks that allow access to almost any job. The dozer blade is used for backfilling, grading, and providing extra stability while digging.

A JD690-B excavator and its bigger counterpart are teamed up on a pipelaying job. Pipeline contractors like the dual capabilities of digging the trench and then craning the pipe sections into position.

Skidders

If agriculture is considered man's first basic industry, surely forestry cannot be far behind. In the past, wood was the primary fuel. Wood remains the main material for housing. Paper, usually made from wood pulp, has been a key element in the spread of knowledge since the Middle Ages.

As mechanization expanded on the farm, the logger adapted some of it to his needs. Many loggers still are farmers with land to clear and wood to sell. Farmers led John Deere into forestry mechanization by using the Lindeman crawler conversion in their woods and orchards.

Logging or tree harvesting is quite diverse. There is no single best way to harvest trees. Originally, trees were cut where nature grew them from natural regeneration. In this mode, the logger must harvest trees where they grow on hillsides, mountains, in swamps or among rocks. He must take the species and size he needs, using the method or system best suited to that particular stand. Machine requirements change when harvesting plantation trees on accessible land. Like farm crops, a single tree species of similar size and spacing is grown in rows for easy harvesting.

John Deere became committed to forestry mechanization in 1965 with the introduction of the 440 skidder. It soon became known as the "loggers dream." Its articulated steering, 45 degrees in both directions, gave good maneuverability on the skid trail. An integral log arch and winch made choker setting fast and easy. A front blade helped clear the trail, and stacked logs at the landing.

Grapple skidders were introduced in the mid '70s and passed the cable units in sales in the late '80s. Larger skidders became the most popular, with the 648D grapple skidder being the best seller in the last half of the '80s.

In 1990 Deere manufactured three grapple skidders and three cable skidders, from 90 to 128 hp. John Deere is the leading manufacturer of forestry equipment in North America.

Here's a JD440-C skidder breaking a new skid trail in the snow. Note the heavily screened operator's station. The left limb riser does double duty as the exhaust stack. It is well protected, and exhaust fumes are blown above the operator's head.

The rugged terrain of the Pacific Northwest presents some real challenges for logging equipment. This 540D, with two sizable logs snugged up to the log arch, had just topped a steep incline on the way to the landing.

Trees grow almost anywhere, even in marshy areas. High-flotation tires help this 548D skidder, with single-function grapple, make it to the landing with a good load. Grapple skidding saves the time spent in choker setting when conditions permit. Hauling cycles go fast when a feller-buncher prebunches the load.

Tire chains are used where tough going demands more tractive effort. This dual-function grapple has a good load of smaller trees. An exclusive Deere feature, constant grapple pressure, assures no loss of load on the haul. The dual-function grapple has fore-and-aft motion as well as lift for easier, faster load building.

Feller-Bunchers and Log Loaders

The chain saw revolutionized the forestry industry by replacing manual chopping and sawing. More recently, the tree shear made it possible to hydraulically force a blade through the tree.

John Deere introduced the JD693 feller-buncher in the mid '70s. This unit clamps the tree, shears it below the clamp, and then places the tree where it can be retrieved. Prebunched loads save the grapple skidder operator considerable cycle time.

The government Soil Bank program created another market for feller-bunchers, to thin trees planted on idle crop land. The feller-buncher, with its ability to cut and bunch rapidly, is the only economical way to harvest these 3-inch-diameter trees.

The John Deere line included four feller-bunchers in 1990, able to cut softwoods up to 20 inches. A choice of saw or shear heads is available. The wheel model is made by CAME-CO Industries Inc., using John Deere engines, axles and other components. The three track models are based on Deere excavators, modified for the woods. Tracks offer high mobility, with good flotation in soft ground and the ability to climb steep slopes. The 693D feller-buncher is able to clear trees 12 to 26 feet from its centerline in a 180-degree path from one setting.

The 693D delimber can remove limbs from trees up to 24 inches in diameter and 50 feet long.

Log loaders work at the landing or sorting yard handling tree-length or cut logs. Holddown clamps provide good control for quick maneuvering. In 1990 John Deere provided five models, from 75 to 160 hp, based on 4-wheel-drive loaders.

The logging industry uses a variety of other equipment from John Deere to clear trails, build haul roads, and prevent or control forest fires. These include crawler dozers, crawler loaders, motor graders, excavators, and scrapers.

Here's the track-mounted 793D feller-buncher clear-cutting on steep terrain. Its long reach and 360-degree swing result in fewer moves and more trees down in prebunched order for fast skidder pickup.

This 643D wheel feller-buncher is a good tool for selective cutting or thinning. It, too, can prebunch for skidder pickup. All John Deere shear heads, either blade or saw, have a safety device. Tree holding clamps must be closed and secure before cutting can start.

The most popular size log loader, the 544E LL, is kept busy sorting precut logs at a sawmill yard. Shown in heavy-duty work, it has smooth-acting hydraulic controls that allow easy placement of individual logs on the saw feed table.

Tree-length loading, usually at a landing or collection point, is an easy job for this 644E log loader. There is plenty of height and hydraulic muscle to top off the largest logging truck.

Self-loading and unloading of trucks is frequently part of the logging system. This 3805 knuckleboom loader is loading out of a stockpile. When the load is complete, the loader will be extended and laid across the top for additional security of the load when transporting.

Other Industrial Equipment

Wherever there is work to be done, someone will figure out the easiest and best way to do it. Frequently the solution is based on John Deere tractor power. Matched equipment does the task, but the tractor furnishes the power, control and mobility.

People like the natural beauty of landscaped areas with grass. Manual labor to reach this goal is reduced by the use of loaders, box scrapers, scarifiers, blades, ditchers, tillers, rakes, seeders and other equipment.

John Deere began offering a variety of grass cutting equipment in the '40s, with the popular Model "LI" industrial tractor and mounted sicklebar mower. Depending on the precision of cut the customer requires, John Deere now sells reel mowers, rotary mowers, flail mowers, sicklebar mowers and rotary cutters. These mowers may be front-mounted, mid-mounted, side-mounted, 3-point hitch or drawn units. They may be powered by the PTO or by hydraulic motors. Most are lifted by the 3-point hitch or a hydraulic cylinder. Most mowing equipment has sufficient sales volume to justify production by John Deere. However, some other mowers and specialty equipment are obtained from smaller short-line companies.

Customers for this varied equipment include individuals, contractors, institutions, the military and government units. State highways and interstate roads use tractors, loaders and rotary cutters.

In 1990 John Deere offered four general purpose tractors, from 55 to 88 net engine horsepower. The base tractor is adequate for many uses. Options permit the customer to tailor a tractor to his needs. All tractors offer good speed selection, with an 8-forward 4-reverse-speed transmission. Mechanical front-wheel drive is an option for difficult traction conditions. PTO options include a continuous or independent rear 540-rpm unit and a 1,000-rpm mid PTO. The 3-point hitch is standard and three remote outlets are available.

John Deere tractors with mowers are common sights wherever grass or weeds are growing. Here is a more unusual application but no real challenge for this combination.

Here's a landscape rake doing a good job of pulverizing and smoothing in a new housing area.

Pipelaying is made easy with a John Deere side boom, a popular option of the '60s. A pipe-bending attachment was available but not shown.

Ornamental stone is beautiful but heavy. The 1010 loader sets it on the truck and later puts it in place at the site. This is just one of the many jobs this unit performs in the landscape business.

Hard-to-reach postholes are just one of the applications for the rotoboom mounted on a JD400 tractor. The loader is one of the most universal tools owned by contractors.

Larger tractors find many uses as prime movers. This JD700 is pulling a sheepsfoot compactor in a fill area. Compaction is a key element in earthmoving projects, and tractor dependability is vital.

What Next?

Meeting Customer Needs

Deere & Company's success in the 1960-90 period was closely related to providing products and services to meet customer needs. The customer base served by John Deere products grew as the company expanded overseas, and as it developed the industrial equipment business and entered the consumer products market. It was no longer entirely a North American farm equipment company.

Planners forecast increasing farm tractor power, based on trends in the '50s. Few foresaw these increases leveling off in the mid '70s, with reduced tillage and lower farm income. The changes in types of equipment used for farming during the 30-year period were dramatic. Labor productivity gains on the farm

BEST SELLING JOHN DEERE FARM EQUIPMENT		
Equipment	**1960**	**1990**
Tractor	730 tricycle	4455
Power	50 hp	140 hp
Primary tillage	810A integral plow	610 chisel plow
Width	4.7 feet	27 feet
Secondary tillage	RW disk	960 field cultivator
Width	13 feet	25.5 feet
Planting	494 planter (plate)	7200 planter (vacuum)
Width	12.7 feet	15 feet
Hay packaging	14T square baler	535 round baler
Bale weight	Up to 70 pounds	Up to 2,000 pounds
Grain harvesting	45 SP combine	9500 SP combine
Platform, width	Rigid, 10 feet	Flex, 20 feet
Corn harvesting	227 corn picker	643 corn head
Width	6.3 feet	15 feet

are likely to continue to outpace those in industry, as they did during this entire period.

A farmer with a new Maximizer combine in 1989 commented, "After I ran that new combine awhile, I invited my wife to ride around the field with me in the buddy seat. As we rode along I told her I never expected to live long enough to have such a good combine." Possibly John Deere customers have had similar thoughts since the advent of the steel plow but never stated them quite so eloquently. Hopefully, customers will be making similar comments on some new John Deere product 30 years from now!

Pipe dream or future reality?

378

PART III

APPENDICES

Deere & Company Corporate Data, 1960

JOHN DEERE FACTORY LOCATIONS
NORTH AMERICA
Des Moines, Iowa
harvesting and cultivating equipment, crop dryers

Dubuque, Iowa
agricultural and industrial tractors, engines

East Moline, Illinois (3)
combine harvesters, windrowers, mowers, hay conditioners; manure spreaders, loaders, feed equipment, forage wagons; foundry

Hoopeston, Illinois
foundry

Horicon, Wisconsin
grain drills, field cultivators, fertilizer distributors

Los Angeles, California
heavy tillage equipment

Moline, Illinois (3)
bulldozers, loaders, backhoes, farm wagons; planters, disks; moldboard plows, bedders, harrows

Ottumwa, Iowa
rakes, balers, rotary choppers, forage harvesters, blowers

Pryor, Oklahoma
nitrogen fertilizers, feed grade urea

Waterloo, Iowa
agricultural tractors, industrial tractors

Welland, Ontario, Canada
rotary cutters, seeding equipment

OVERSEAS
Mannheim, Germany
agricultural tractors

Monterrey, Mexico
agricultural tractors, farm wagons, tillage equipment

Rosario, Argentina
agricultural tractors

Zweibrücken, Germany
combine harvesters, balers

ASSOCIATED COMPANIES
SOCIETE REMY ET FILS
Senonches, France
cultivating equipment, hay tools, forage harvesters

ETS. R. ROUSSEAU
Orleans, France
low-density balers, threshing machines

STE. THIEBAUD-BOURGUIGNONNE
Arc-les-Gray, France
hay tools, grain separators

MARKETING UNITS
UNITED STATES SALES BRANCHES
Atlanta, Georgia
Baltimore, Maryland
Columbus, Ohio
Dallas, Texas
Indianapolis, Indiana
Kansas City, Missouri
Lansing, Michigan
Minneapolis, Minnesota
Moline, Illinois
Omaha, Nebraska
Portland, Oregon
San Francisco, California
St. Louis, Missouri
Syracuse, New York

CANADIAN SALES BRANCHES
Calgary, Alberta
Hamilton, Ontario
Regina, Saskatchewan
Winnipeg, Manitoba

OVERSEAS SALES BRANCHES
Mannheim, Germany
Mexico City, Mexico
Sydney, Australia

EXPORT SALES BRANCHES
Mannheim, Germany
Moline, Illinois

Deere & Company Corporate Data, 1990

JOHN DEERE FACTORY LOCATIONS

NORTH AMERICA

Coffeyville, Kansas
power transmission equipment

Davenport, Iowa
construction equipment

Des Moines, Iowa
cotton harvesting equipment, tillage and planting equipment

Dubuque, Iowa
construction and forestry equipment, engines

East Moline, Illinois (2)
combine harvesters; foundry

Greeneville, Tennessee
residential lawn care products/ walk-behind mowers

Horicon, Wisconsin
lawn and grounds care and outdoor power equipment

Moline, Illinois
planting equipment and hydraulic cylinders

Ottumwa, Iowa
hay and forage equipment, golf and turf equipment

Waterloo, Iowa (5)
agricultural tractors; product engineering; foundry; engines and other major components

Welland, Ontario, Canada
material handling equipment, rotary cutters, utility vehicles

Wood-Ridge, New Jersey
rotary engines

OVERSEAS

Arc-les-Gray, France
balers, forage equipment

Bruchsal, Germany
tractor and combine cabs

Madrid, Spain
agricultural tractors and components

Mannheim, Germany
agricultural tractors

Nigel, South Africa
agricultural tractors, tillage equipment

Rosario, Argentina
agricultural tractors, engines

Saran, France
engines

Welshpool, Australia
tillage equipment

Zweibrücken, Germany
combine harvesters

AFFILIATED COMPANIES

DEERE-HITACHI CONSTRUCTION MACHINERY CORPORATION

Kernersville, North Carolina
hydraulic excavators

JOHN DEERE S.A. DE C. V. (MEXICO)

Monterrey, Mexico
tillage tools, cultivating equipment

Saltillo, Mexico
agricultural tractors, engine assembly

SLC S.A. INDUSTRIA E COMERCIO

Horizontina, Brazil
combine harvesters

MARKETING UNITS

FARM EQUIPMENT SALES BRANCHES

North America
Atlanta, Georgia
Columbus, Ohio
Dallas, Texas
Kansas City, Missouri
Minneapolis, Minnesota
Raleigh-Durham, North Carolina
Grimsby, Ontario, Canada

Overseas
Arlöv, Sweden
Madrid, Spain
Mannheim, Germany
Milan, Italy
Nigel, South Africa
Nottingham, England
Ormes, France
Rosario, Argentina
Welshpool, Australia

INDUSTRIAL REGION SALES OFFICES

Baltimore, Maryland
Denver, Colorado
Moline, Illinois
Grimsby, Ontario, Canada

EXPORT SALES BRANCHES

Mannheim, Germany
Moline, Illinois

PARTS DISTRIBUTION CENTERS

Milan, Illinois
Bruchsal, Germany

SPECIAL SALES DIVISIONS

MOLINE, ILLINOIS

Deere Marketing Services, Inc.
John Deere Catalog Company
John Deere Merchandise (JDM)
National Sales Division
Power Systems Group

FINANCIAL SERVICES SUBSIDIARIES

JOHN DEERE CREDIT COMPANY

MOLINE, ILLINOIS
John Deere Capital Corporation
Reno, Nevada
Deere Credit, Inc.
Moline, Illinois
Deere Credit Services, Inc.
Des Moines, Iowa
Farm Plan Corporation
Madison, Wisconsin
John Deere Leasing Company
Moline, Illinois

JOHN DEERE FINANCE LIMITED

EDMONTON, ALBERTA, CANADA

JOHN DEERE INSURANCE GROUP

MOLINE, ILLINOIS
John Deere Insurance Company
Moline, Illinois
John Deere Life Insurance Company
Jacksonville, Illinois

JOHN DEERE INSURANCE COMPANY OF CANADA

GRIMSBY, ONTARIO, CANADA

HEALTH MANAGEMENT SERVICES

John Deere Health Care, Inc.,
Moline, Illinois

U.S. Tractor Specifications
Dubuque, Mannheim, and Yanmar Diesel Tractors

Model	Trans-mission		Engine				Wheel-base (in.)	Rear Tires	Hitch Lift (lb.)	Sound Level dB(A)	Last List Price
		Cylin-ders	Bore & Stroke	Disp. (cu. in.)	Rated rpm						
Yanmar											
1250	9	SG	3	3.74x4.33	143	2500	75	14.9-28	2240	94.5	$14,220
1450	9	SG	4	3.74x4.33	190	2400	83	16.9-28	2652	95.5	$16,234
1650	9	SG	4T	3.74x4.33	190	2300	83	18.4-26	2698	92.5	$18,382
Dubuque											
1010	5	SG	4	3.62x3.50	144	2500	78	12.4-28			$3,722
2010	8	SR	4	3.88x3.50	165	2500	87	13.6-28			$4,534
Dubuque and Mannheim											
830	8	CS	3	3.86x4.33	152	2400	74	13.6-28	1780	97.0	$6,111
1020	8	CS	3	3.86x4.33	152	2500	74	13.6-28	2010		$4,485
2040	8	CS	3	4.02x4.33	164	2500	74	14.9-28	1825	96.5	$13,970
2150	8	CS	3	4.19x4.33	179	2500	75	16.9-28	2023	93.0	$16,731
2155	8	CS	3	4.19x4.33	179	2500	81	16.9-28	2584	94.0	$17,177
1520	8	CS	3	4.02x4.33	164	2500	74	14.9-28	2300		$5,248
1530	8	CS	3	4.02x4.33	164	2500	74	14.9-28	2010	98.5	$7,145
2240	8	CS	3	4.19x4.33	179	2500	74	14.9-28	2243	98.0	$15,476
2350	8	CS	4	4.19x4.33	239	2500	89	16.9-30	2736	93.5	$19,543
2355	8	CS	4	4.19x4.33	239	2500	89	16.9-30	3438	93.5	$19,994
2020	8	CS	4	3.86x4.33	203	2500	86	14.9-28			$5,788
2030	8	CS	4	4.02x4.33	219	2500	86	16.9-28	2300	97.0	$6,235
2440	8	CS	4	4.02x4.33	219	2500	86	16.9-28	2250	97.0	$18,583
2550	8	CS	4	4.19x4.33	239	2500	89	18.4-30	2736	95.0	$21,813
2555	8	CS	4	4.19x4.33	239	2500	89	18.4-30	3439	97.0	$23,280
2630	8	CS	4	4.19x5.00	276	2500	86	16.9-28	2300	97.0	$10,087
2640	8	CS	4	4.19x5.00	276	2500	86	16.9-28	2250	97.0	$21,150
2750	8	CS	4T	4.19x4.33	239	2500	89	18.4-30	3372	93.5	$24,409
2755	8	CS	4T	4.19x4.33	239	2500	89	18.4-30	3720	76.5 cab	$34,561
Mannheim											
2840	12	CS	6	4.02x4.33	329	2500	97	18.4-34	3935	94.5	$16,971
2940	16	TSS	6	4.19x4.33	359	2500	100	18.4-34	3884	93.0	$25,117
2950	16	TSS	6	4.19x4.33	359	2500	100	18.4-38	4868	76.0 cab	$34,699
2955	16	TSS	6	4.19x4.33	359	2300	100	18.4-38	5235	77.0 cab	$38,981
3150	16	TSS		4.19x4.33	359	2400	102	18.4-38	6300	77.5 cab	$43,000
3155	16	TSS		4.19x4.33	359	2400	102	18.4-38	6308		

Notes: Transmission lists number of forward speeds and type: SG sliding gear, CS collar shift, TSS top shaft synchronized or SR Syncro-Range.
Engine cylinders show number and T if turbocharged. Sound level is at the operator's ear at 75% load.

Nebraska Tractor Tests
Dubuque, Mannheim, and Yanmar Diesel Tractors

Model	Year	Test No.	Max. PTO hp	Drawbar			Max. Pull (lb.)	Weight with Ballast (lb.)	Fuel Use at:	
				Maximum hp at mph		Lugging Increase (%)			Max. PTO hp at Std PTO rpm (hp-hr/gal.)	75% of Pull at Max. Power (hp-hr/gal.)
Yanmar										
1250	1982	1437	40.7	34.0	5.4	17	3750	5180	17.0	12.7
1450	1983	1505	51.4	43.8	7.7	20	5321	6570	17.6	13.7
1650	1983	1506	62.2	54.1	7.8	18	6572	7785	18.8	15.0
Dubuque										
1010	1961	803	36.0	30.8	6.5		3657	5754	12.6	10.7
2010	1961	799	46.7	41.4	6.3		4553	6392	13.2	10.5
Dubuque and Mannheim										
830	1973	1146	35.3	28.5	5.2	22	4252	5714	14.7	11.0
1020	1966	937	38.9	32.3	5.4	27	4200	5955	15.5	11.2
2040	1975	1191	40.9	34.2	5.6	14	4634	6420	15.5	11.1
2150	1983	1469	46.5	38.7	7.0	16	4875	5930	14.8	11.2
2155	1987	024	45.6	37.4	7.7		3975	5300	15.3	11.7
1520	1968	991	46.5	37.6	4.0	18	5082	6970	16.0	12.0
1530	1973	1147	45.4	38.9	5.6	27	5139	6863	14.3	11.5
2240	1975	1192	50.4	41.3	5.5	17	5433	7340	15.2	11.2
2350	1983	1470	56.2	47.3	6.8	14	5688	7190	15.7	11.9
2355	1987	025	55.9	46.7	8.2		6195	6965	16.3	12.6
2020	1966	938	54.1	47.4	5.6	19	5547	7310	16.4	12.1
2030	1971	1085	60.6	52.3	5.9	21	6472	7430	16.3	11.8
2440	*Used 2030 test No. 1085*									
2550	1983	1471	65.9	56.2	6.2	14	6979	8485	15.4	12.1
2555	1987	027	66.0	56.5	5.7		6910	7300	16.4	13.2
2630	1974	1157	70.4	58.2	5.4		6470	8430	16.0	11.0
2640	*Used 2630 test No. 1157*									
2750	1983	1472	75.4	64.0	4.8	14	7581	9625	16.1	12.4
2755	1986	1605	76.6	66.0	5.2	22	7498	9715	17.8	14.2
Mannheim										
2840	1977	1249	80.6	67.5	6.2	14	8386	10500	16.0	11.6
2940	1980	1351	81.2	68.0	4.6	15	7601	10650	15.9	11.9
2950	1983	1473	85.4	73.3	6.6	15	8768	11045	16.3	12.4
2955	1986	1606	86.2	74.2	5.4	24	9971	11150	17.4	13.8
3150	1986	1589	96.1	83.0	5.6	20	10507	12300	16.4	13.3
3155	*Used 3150 test No. 1589*									

Sources: *Nebraska Tractor Test Data*
Implement & Tractor Red Book
Official Guide, Tractors and Farm Equipment

383

U.S. Tractor Specifications
Waterloo Diesel Tractors

Model	Trans-mission	Cylin-ders	Bore & Stroke	Disp. (cu. in.)	Rated rpm	Wheel-base (in.)	Rear Tires	Hitch Lift (lb.)	Sound Level dB(A)	Last List Price
8010	9	6	4.50x5.00	425	2100	120	23.1-26			
2510	8 SR	4	3.86x4.33	203	2500	90	13.6-38			$5,879
2520	8 SR	4	4.02x4.33	219	2500	90	13.6-38			$7,546
3010	8 SR	4	4.12x4.75	254	2200	90	13.6-38			$5,362
3020	8 SR	4	4.25x4.75	270	2500	90	15.5-38			$8,351
4030	8 SR	6	4.02x4.33	329	2500	101	16.9-34	3520	83.5 cab	$17,721
4040	16 QR	6	4.25x4.75	404	2200	104	18.4-34	4316	77.5 cab	$35,229
4050	16 QR	6	4.57x4.75	466	2200	107	18.4-34D	6294	74.0 cab	$44,670
4055	16 QR	6T	4.57x4.75	466	2200	107	18.4-38D	6550	76.0 cab	$48,928
4010	8 SR	6	4.12x4.75	380	2200	96	16.8-34			$6,167
4020	8 SR	6	4.25x4.75	404	2200	98	18.4-34	3790		$10,345
4000	8 SR	6	4.25x4.75	404	2200	96	16.9-34			$9,389
4230	8 SR	6	4.25x4.75	404	2200	104	20.8-34	4350	82.5 cab	$20,342
4240	16 QR	6	4.57x4.75	466	2200	107	18.4-34	5732	79.0 cab	$39,606
4250	16 QR	6T	4.57x4.75	466	2200	107	18.4-38D	6294	73.5 cab	$49,080
4255	16 QR	6T	4.57x4.75	466	2200	107	18.4-38D	6550	76.5 cab	$53,488
4320	8 SR	6T	4.25x4.75	404	2200	106	18.4-38	3790	92.5 cab	$11,312
4430	8 SR	6T	4.25x4.75	404	2200	106	18.4-38	4440	82.5 cab	$22,451
4440	16 QR	6T	4.57x4.75	466	2200	107	18.4-38	5732	78.0 cab	$43,634
4450	16 QR	6T	4.57x4.75	466	2200	107	18.4-38D	6294	73.5 cab	$52,409
4455	16 QR	6T	4.57x4.75	466	2200	107	18.4-42D	6550	76.0 cab	$57,106
4555	16 QR	6T	4.57x4.75	466	2200	118	18.4-42D	8870	76.0 cab	$61,450
4520	8 SR	6T	4.25x4.75	404	2200	106	20.8-38			$11,600
4620	8 SR	6TA	4.25x4.75	404	2200	106	20.8-38	5700	93.0 cab	$13,286
4630	8 SR	6TA	4.25x4.75	404	2200	112	18.4-38	5260	83.0 cab	$27,076
4640	16 QR	6TA	4.57x4.75	466	2200	118	18.4-42	7328	77.5 cab	$51,553
4650	16 QR	6TA	4.57x4.75	466	2200	118	20.8-38D	8475	73.5 cab	$61,264
4755	16 QR	6TA	4.57x4.75	466	2200	118	20.8-42D	8870	76.0 cab	$69,882
4840	8 PS	6TA	4.57x4.75	466	2200	118	20.8-38	8295	79.0 cab	$57,648
4850	15 PS	6TA	4.57x4.75	466	2200	118	20.8-38D	9599	75.0 cab	$70,800
4955	15 PS	6TA	4.57x4.75	466	2200	118	20.8-42D	9710	76.5 cab	$80,225
5010	8 SR	6	4.75x5.00	531	2200	104	24.5-32			$10,733
5020	8 SR	6	4.75x5.00	531	2200	104	24.5-32	5110		$14,550
6030	8 SR	6TA	4.75x5.00	531	2100	104	20.8-38	5910	86.5 cab	$28,745
7020	8 SR	6TA	4.25x4.75	404	2200	120	18.4-34	5700	96.0 cab	$21,724
8430	16 QR	6TA	4.57x4.75	466	2100	125	20.8-34	6400	80.5 cab	$43,144
8440	16 QR	6TA	4.57x4.75	466	2100	125	20.8-38D	8545	79.0 cab	$63,518
8450	16 QR	6TA	4.57x4.75	466	2100	125	20.8-38D	8545	77.0 cab	$74,296
8560	24 P	6TA	4.57x4.75	466	2100	134	18.4-42D	13940	76.5 cab	$100,590
7520	8 SR	6TA	4.75x5.00	531	2100	120	23.1-30		87.5 cab	$25,189
8630	16 QR	6TA	5.12x5.00	619	2100	125	23.1-30	7300	82.5 cab	$52,475
8640	16 QR	6TA	5.12x5.00	619	2100	125	23.1-34D	8545	80.0 cab	$77,628
8650	16 QR	6TA	5.12x5.00	619	2100	125	23.1-34D	8545	76.5 cab	$93,151
8760	24 P	6TA	5.12x5.00	619	2100	134	18.4-42D	13940	76.0 cab	$117,390
8850	16 QR	V8TA	5.51x5.00	955	2100	133	24.5-32D	10118	78.0 cab	$118,609
8960	24 P	6TA	5.50x6.00	855	1900	134	20.8-42D	13940	74.0 cab	$137,000

Notes: Transmission lists number of forward speeds and type: SR Syncro-Range, QR Quad-Range, PS Power Shift or P PowrSync. Engine cylinders show number and aspiration: T Turbocharged or TA turbocharged and aftercooled. Rear tires show size of test tires followed by "D" if duals were used.

Nebraska Tractor Tests
Waterloo Diesel Tractors

Model	Year	Test No.	Max. PTO hp	Maximum hp at mph	Lugging Increase (%)	Max. Pull (lb.)	Weight with Ballast (lb.)	Max. PTO hp at Std PTO rpm (hp-hr/gal.)	75% of Pull at Max. Power (hp-hr/gal.)
				Drawbar				**Fuel Use at:**	
8010	*No Nebraska Test*			150 est.			24860		
2510	1965	916	55.0	48.9 4.4		6761	8485	16.4	13.3
2520	1968	992	61.3	54.7 5.4	19	6650	8955	15.3	12.4
3010	1960	762	59.4	54.5 5.8		6323	8640	15.5	13.0
3020	1963	848	65.3	57.1 5.0	16	7536	9585	13.8	10.6
4030	1972	1111	80.3	68.8 6.1	32	7512	10150	15.4	11.0
4040	1977	1267	90.8	77.8 6.4	12	8936	12055	13.5	11.0
4050	1983	1474	101.5	92.0 6.5	14	10331	13005	15.1	12.5
4055	1989	064	108.7	99.1 4.3	32	13378	13660	16.1	13.4
4010	1960	761	84.0	73.6 5.7		7002	9775	15.7	12.4
4020	1963	849	91.2	78.0 4.9		10184	13055	14.9	11.8
4000	1969	1023	96.9	82.6 5.7	13	8215	10870	15.6	13.1
4230	1972	1112	100.3	86.4 6.6	16	11116	14150	14.1	11.9
4240	1977	1266	110.9	95.2 5.4	16	11492	14090	14.3	11.4
4250	1983	1475	120.2	108.8 5.7	27	12858	14990	15.7	12.8
4255	1989	065	123.7	111.9 6.0	33	14635	14805	16.5	13.8
4320	1970	1050	116.6	104.5 6.4	14	11783	14380	15.3	11.8
4430	1972	1110	125.9	108.5 6.6	22	12032	15060	15.6	12.0
4440	1977	1265	130.6	112.6 5.0	15	13392	15540	15.2	11.8
4450	1983	1476	140.3	125.2 5.0	24	15069	16680	16.4	13.4
4455	1989	066	142.7	128.8 5.6	40	17161	17210	17.0	14.1
4555	1989	067	156.8	141.8 5.7	37	17774	18745	17.4	145.5
4520	1969	1015	123.4	111.2 5.1	6	13221	17850	15.5	12.8
4620	1971	1073	135.8	117.4 5.1	16	13916	18640	15.6	11.9
4630	1972	1113	150.7	134.6 10.3	23	14478	16250	15.7	12.9
4640	1977	1264	156.3	134.7 6.4	21	15362	18120	15.6	13.2
4650	1983	1477	165.7	145.6 6.4	23	16814	19615	17.0	13.8
4755	1989	068	177.1	156.0 7.1	37	19898	20320	18.2	15.4
4840	1977	1263	180.6	157.1 4.8	28	17307	20370	15.8	12.5
4850	1982	1461	193.0	171.3 8.0	26	18789	21435	17.5	13.8
4955	1989	060	202.7	173.5 3.9	40	23742	24665	18.4	14.6
5010	1962	828	121.1	108.9 5.1		14174	17175	15.5	12.9
5020	1966	947	133.2	116.4 6.3		16197	21360	16.9	13.5
6030	1972	1100	176.0	155.4 6.6	21	15303	18180	15.8	12.7
7020	1971	1063	146.2	131.5 5.3	18	18726	19570	14.3	11.8
8430	1975	1179	178.2	160.4 5.6	24	22503	23100	15.1	12.6
8440	1979	1323	179.8	163.3 5.0	20	25383	26590	15.9	13.1
8450	1982	1436	187.0	175.7 4.5	24	26567	29150	15.7	13.5
8560	1989	061	202.6	180.7 6.1	39	32166	32075	17.2	14.1
7520	1972	1101	175.8	165.2 5.7	18	22517	22320	15.2	12.8
8630	1975	1180	225.6	202.3 4.6	23	26443	26480	15.3	12.8
8640	1979	1324	228.8	208.4 4.6	26	27782	28270	15.6	13.1
8650	1982	1435	238.6	221.8 5.6	21	28257	30270	15.5	13.2
8760	1989	062	256.9	240.2 4.6	41	32880	32695	16.3	14.0
8850	1982	1434	304.0	274.2 5.2	29	35330	37700	15.7	13.3
8960	1989	063	333.4	308.2 5.9	38	34316	35570	17.1	14.3

Sources: *Nebraska Tractor Test Data*
Implement & Tractor Red Book
Official Guide, Tractors and Farm Equipment

Worldwide John Deere Tractor Comparisons

Year	United States Model	PTO hp	Mannheim Model	Engine SAE hp*	Cylinders	Notes	Others Model	Engine hp	Country
1960-65									
			300	36	4	Lanz 10-speed	303	37	France
	1010	36	500	42	4	Lanz 10-speed	505	44	France
	2010	46	700	53	4	Lanz 10-speed	505	44	Spain
			100	22	2	Lanz 10-speed	445	36	Argentina
			200	28	2	Lanz 10-speed			
1966-71									
			310	36	3	Lanz 10-speed	515	45	Spain
	1020	38	510	45	3	Lanz 10-speed	717	56	Spain
	2020	54	710	56	4	Lanz 10-speed	818	60	Spain
							1420	43	Argentina
	820	31	820	34	3	U.S. 8-speed	2420	66	Argentina
			920	40	3	U.S. 8-speed	3420	77	Argentina
	1520	46	1020	47	3	U.S. 8-speed	4420	102	Argentina
			1120	52	3	U.S. 8-speed	1020	45	Spain
			2020	64	4	U.S. 8-speed	1520	52	Spain
			2120	72	4	U.S. 8-speed	2020	61	Spain
			3120	81	6		2120	68	Spain
							3120	82	Spain
1972-75									
							1530	48	Spain
			1630	59	3		1630	57	Spain
	2030	60	2030	71	4		2030	68	Spain
			2130	79	4		2130	75	Spain
			3130	97	6		3130	90	Spain
							4235	100	Mexico
	830	35					4435	125	Mexico
	1530	45					2535	60	Mexico
	2630	70					2735	71	Mexico
			New Styling				2330	52	Argentina
			830	35	3		2530	67	Argentina
			930	43	3		2730	78	Argentina
			1030	48	3		3530	101	Argentina
			1130	57	3		4530	114	Argentina
			1830	71	4		3380	68 PTO	Australia
							4080	85 PTO	Australia
							4280	98 PTO	Australia
							4480	119 PTO	Australia

* Farm tractors sold in the U.S. are rated by PTO horsepower. Most other countries measure tractor power by engine hp, which is about 20% more. Mannheim tractors sold in England are rated by SAE hp; those sold in Germany are rated by PS, the German term for horsepower.

Worldwide John Deere Tractor Comparisons

Year	United States		Mannheim				Others		
	Model	PTO hp	Model	Engine PS*	Cylinders	Notes	Model	Engine hp	Country
1976-82									
	2040	40	1030	48	3		3330	78	Argentina
	2240	50	1630	59	3		1035	48	Spain
	2440	60	1830	71	4		1635	57	Spain
	2640	70	2130	79	4		2035	68	Spain
	2840	80	3130	97	6		2135	75	Spain
			830	35	3		3135	90	Spain
			3030	86	6				
			840	38	3		1040SV	54	Spain
			940	44	3		1140SF	60	Spain
			1040	50	3		1840SM	72	Spain
			1140	56	3		2040S-2	80	Spain
			1640	62	4		2140S-4	91	Spain
			2040	70	4		3140S-2	104	Spain
			2040S	75	4		3340S-4	110	Spain
			2140	82	4		3640	126	Spain
			3040	90	6		2140	90	Argentina
	2940	81	3140	97	6		3140	96	Argentina
			3640-MFWD	112	6	MFWD standard	3540	121	Argentina
	4040	90	4040S	115	6		4040	103	Argentina
	4240	110	4240S	132	6				
1983-86									
			4350	140	6		1641	55	S. Africa
			1350	38	3	MFWD available	2141	68	S. Africa
			1550	44	3	MFWD available	2541	76	S. Africa
			1750	50	3	MFWD available	2941	85	S. Africa
	2150	46	1850	56	3	MFWD available	3141	88	S. Africa
	2350	56	2250	62	4	MFWD available	2555	72	Mexico
	2550	66	2450	70	4	MFWD available	2755	82	Mexico
	2750	75	2650	78	4T	Front hitch & PTO	4255	140	Mexico
			2850	86	4T	Front hitch & PTO	4455	153	Mexico
	2950	85	3050	92	6	Front hitch & PTO	4090	94 PTO	Australia
	3150	96	3350	100	6	Front hitch & PTO	4290	110 PTO	Australia
			3650-MFWD	114	6T	Front hitch & PTO	4490	129 PTO	Australia
						Front hitch & PTO	4690	154 PTO	Australia
						Front hitch & PTO	3540	121	Argentina
1987-90									
			1550	46	3	MFWD available	2251	61	S. Africa
	2155	46	1850	56	3	MFWD available	2251N	69	S. Africa
	2355	56	2250	62	4	MFWD available	2351	74	S. Africa
	2555	65	2450	70	4	MFWD available	2651	80	S. Africa
	2755	75	2650	78	4T	MFWD available	2951	96	S. Africa
	2955	85	3050	92	6	MFWD available	3351	99	S. Africa
			3350	100	6	MFWD available	3651	110	S. Africa
			3650-MFWD	114	6T	MFWD standard	2850	95	Argentina
			1950	62	3T	MFWD available			
	3155	95					3350	110	Argentina
	4055	105	4055-MFWD	128	6T	MFWD standard	3550	125	Argentina
	4255	120	4255-MFWD	144	6T	MFWD standard			
	4455	140	4455-MFWD	160	6T	MFWD standard			
	4555	155							
	4755	175	4755-MFWD	190	6TA	MFWD standard			
	4955	200	4955-MFWD	228	6TA	MFWD standard			

Serial Numbers* of U.S. Farm Tractors
Made by Dubuque, Mannheim, or Yanmar

820
Serial	Year
10000	1968
23100	1969
36000	1970
54000	1971
71850	1972
90200	1973

830
Serial	Year
108507	1974
155914	1975

1010
Serial	Year
10001	1961
23630	1962
32188	1963
43900	1964
53722	1965

1020
Serial	Year
14501	1965
14682	1966
42715	1967
65184	1968
82409	1969
102038	1970
117500	1971
134700	1972
157109	1973

1250
Serial	Year
1000	1982
1256	1983
3001	1984
4001	1985
5001	1986
5501	1987

1450
Serial	Year
1000	1984
2201	1985
3001	1986
3501	1987

1520
Serial	Year
76112	1968
82405	1969
102061	1970
117500	1971
134700	1972
157109	1973

1530
Serial	Year
176601T	1974
108811L	
145500L	1975

1650
Serial	Year
1000	1984
2401	1985
3001	1986
3501	1987

2010
Serial	Year
10001	1961
21087	1962
31250	1963
44036	1964
58186	1965

2020
Serial	Year
14502	1965
14680	1966
42721	1967
65176	1968
82404	1969
102032	1970
117500	1971

2030
Serial	Year
134700	1972
157109	1973
187301T	1974
140000L	
213350T	1975
145500L	

2040
Serial	Year
179963	1976
221555	1977
266057	1978
304165	1979
336935	1980
392026	1981
419145	1982

2150
Serial	Year
433467	1983
505001	1984
532000	1985
562001	1986

2155
Serial	Year
600000	1987
624800	1988
652073	1989
686146	1990

2240
Serial	Year
179298	1976
221716	1977
277267	1978
305307	1979
337767	1980
392292	1981
418608	1982

2350
Serial	Year
433474	1983
505001	1984
532000	1985
562001	1986

2355
Serial	Year
600000	1987
624800	1988
653366	1989
685855	1990

2355N
Serial	Year
600000	1987
624800	1988
654199	1989
685855	1990

2440
Serial	Year
235210	1976
258106	1977
280789	1978
305501	1979
335625	1980
362173	1981
376746	1982

2550
Serial	Year
433480	1983
505001	1984
532000	1985
562001	1986

2555
Serial	Year
600000	1987
624800	1988
651786	1989
685748	1990

2630
Serial	Year
188601	1974
213360	1975

2640
Serial	Year
235313	1976
258106	1977
280789	1978
305505	1979
335628	1980
362175	1981
376744	1982

2750
Serial	Year
433494	1983
505001	1984
532000	1985
562001	1986

2755
Serial	Year
600000	1987
624800	1988
650392	1989
685854	1990

2840
Serial	Year
214909	1977
264711	1978
304654	1979

2855N
Serial	Year
600000	1987
624800	1988
652661	1989
685906	1990

2940
Serial	Year
350586	1980
390496	1981
418953	1982

2950
Serial	Year
433508	1983
505001	1984
532000	1985
562001	1986

2955
Serial	Year
600000	1987
624800	1988
652968	1989
685843	1990

3150
Serial	Year
532000	1985
562001	1986
587950	1987

3155
Serial	Year
618645	1987
624591	1988
653652	1989
685845	1990

* Beginning serial numbers are given for John Deere farm tractors sold in the United States from 1960 through 1990. The model year usually starts August 1 of the preceding calendar year and continues through July 31 of the year shown. The model years for the 1020, 1520, and 2020 were November 1 through October 31.

Serial Numbers of U.S. Farm Tractors

Made by Waterloo

2510	
1000	1966
8958	1967
14291	1968

2520	
17000	1969
19416	1970
22000	1971
22911	1972
23865	1973

3010	
1000	1961
10801	1962
32400	1963

3020	
50000	1964
68000	1965
84000	1966
97286	1967
112933	1968
123000	1969
129897	1970
150000	1971
154197	1972

4000	
211422	1969
222143	1970
250000	1971
260791	1972

4010	
1000	1961
20200	1962
38200	1963

4020	
65000	1964
91000	1965
119000	1966
145660	1967
173982	1968
201000	1969
222143	1970
250000	1971
260791	1972

4030	
1000	1973
6700	1974
10153	1975
13022	1976
15417	1977

4040	
1000	1978
3199	1979
6033	1980
8707	1981
11727	1982

4050	
1000	1983
3501	1984
5001	1985
6501	1986
7001	1987
7501	1988

4055	
1001	1989
2501	1990

4230	
1000	1973
13000	1974
22074	1975
28957	1976
35588	1977

4240	
1000	1978
7434	1979
14394	1980
20186	1981
25670	1982

4250	
1000	1983
6001	1984
9001	1985
11001	1986
12501	1987
13501	1988

4255	
1001	1989
3001	1990

4320	
6000	1971
17031	1972

4430	
1000	1973
17500	1974
33050	1975
47222	1976
62960	1977

4440	
1000	1978
14820	1979
29539	1980
42665	1981
56346	1982

4450	
1000	1983
11001	1984
18001	1985
22001	1986
24001	1987
26001	1988

4455	
1001	1989
5001	1990

4520	
1000	1969
7005	1970

4555	
1001	1989
3001	1990

4620	
10000	1971
13692	1972

4630	
1000	1973
7022	1974
11717	1975
18392	1976
25794	1977

4640	
1000	1978
7422	1979
13860	1980
19459	1981
25729	1982

4650	
1000	1983
7001	1984
10001	1985
12501	1986
14001	1987
15501	1988

4755	
1001	1989
3001	1990

4840	
1000	1978
4233	1979
7539	1980
11042	1981
14933	1982

4850	
1000	1983
5001	1984
8001	1985
10001	1986
11001	1987
12001	1988

4955	
1001	1989
3501	1990

5010	
1000	1963
4500	1964
8000	1965

5020	
12000	1966
15650	1967
20399	1968
24038	1969
26624	1970
30000	1971
30608	1972

6030	
33000	1972
33550	1973
34586	1974
35400	1975
36014	1976
36577	1977

7020	
1000	1971
2006	1972
2700	1973
3156	1974
3579	1975

7520	
1000	1972
1600	1973
3054	1974
4945	1975

8010	
1000	1961

8020	
1000	1964

8430	
1000	1975
1690	1976
3962	1977
5323	1978

8440	
1001	1979
2266	1980
3758	1981
5235	1982

8450	
1000	1982
2000	1983
3501	1984
5001	1985
5501	1986
6001	1987
6501	1988

8560	
1001	1989
1501	1990

8630	
1000	1975
2382	1976
5222	1977
7626	1978

8640	
1500	1979
3198	1980
5704	1981
7960	1982

8650	
1500	1982
3000	1983
5001	1984
7001	1985
8001	1986
8501	1987
9001	1988

8760	
1001	1989
2001	1990

8850	
2000	1982
4000	1983
5101	1984
6001	1985
6501	1986
7001	1987
7501	1988

8960	
1001	1989
1501	1990

Model	SP/PTO	Platform Width (ft)	Corn Head Rows	Cylinder Dimensions (in.)	Cylinder Speed Range (rpm)	Straw Walkers	Separation/Cleaning Area (sq. in.)	Grain Tank (bu)	Engine hp (SAE)
Hi-Lo									
40	SP	8-10'	2	22 x 24.675	394-1075	3	2710/1897	35	42
42	PTO	9'	2	22 x 24.675	394-1075	3	2710/1897	35	--
45	SP	8-12'	2	22 x 26	394-1075	3	3380/2266	50	59
45R									
55	SP	12-15'	2/3	22 x 30	196-1190	3	4200/3007	60/65	72
55H									
55R									
65	PTO	12' & PU Platform		22 x 30	196-1190	3	4200/3007	55	71
95	SP	12-20'	2/3/4	22 x 40	196-1190	4	5600/4061	60/80	80/90
95H								80	90
95R									
105	SP	12-22'	3/4/6	22 x 50	276-1190	5	6930/5066	75/100	105
105R	SP	12-22'		22 x 50	276-1190	5	6930/5066	75/100	105
New Generation									
3300	SP	10-15'	2/3	22 x 28.75	387-1172	3	3448/2484	59	70
4400	SP	10-20'	2/3/4	22 x 38	387-1172	4	4559/3357	102	95
4400R									
6600	SP	13-22'	3/4/6	22 x 44	407-1229	4	5728/4478	112	104
6600R									
6601	PTO	13' & PU Platform		22 x 44	407-1229	4	5728/4478	100	---
6602	SP	18-20'		22 x 44	350-1058	4	5728/4478	112	121
7700	SP	13-24'	4/6/8	22 x 55	407-1229	5	7124/5637	129	128
7700R									128/145
Titan									
4420	SP	13-20'	2/3/4	22 x 38	390-1170	4	4559/3357	102	100
4420R									
6620	SP		4/6	22 x 44	430-1230	4	6593/4478	166	120/145
6620R									
6622	SP	18-20'		22 x 44	370-1030	4	6593/4478	122	145
7720	SP	13-30'	4/5/6	22 x 55	430-1230	5	8222/5637	190	145/165
7720R									
7721	PTO	12-14' & PU Platform		22 x 55	430-1230	5	8222/5637	190	---
8820	SP	15-22'	5/6/8/12	22 x 65.5	430-1230	6	9867/6799	222	200
8820R									
Titan II									
6620	SP	13-24'	4/6	22 x 44	350-1240	4	6593/4887	166	125/145
SH6620		13-22'	4/6						145
6620R		13-24'							125/145
6622	SP	22'		22 x 44	350-1240	4	6593/4887	122	145
7720	SP	13-30'	4/5/6	22 x 55	350-1240	5	8222/6179	190	145/165
7720R									
7722H	SP	22-24'		22 x 55	350-1240	5	8222/6179	160	225
8820	SP	15-30'	5/6/8/12	22 x 65.5	350-1240	6	9867/7498	222	225
8820R									
4425	SP	13-18'	2/3/4	24 x 41	380-1100	4	5890/5626	125	117
Maximizer									
9400	SP	13-30'	4/5/6	26 x 53.5	240-970	4	8618/6370	182	167
9400R		18' Draper							
9500	SP	13-30'	4/5/6/8	26 x 53.5	240-970	4	9703/6370	204	200
9500R		18' Draper							
9501	PTO	12-14' & PU Platform		26 x 53.5	240-970	4	9703/6370	204	---
9600	SP	13-35'	5/6/8/12	26 x 64.5	240-970	5	11625/7719	240	200/253
9600R		18' Draper							
4435	SP	13-18'	4	24 x 41	380-1100	4	6290/5626	125	117

Combine Specifications — Overseas Models

Model	SP/PTO	Platform Width (ft)	Corn Head Rows	Cylinder Dimensions (in.)	Cylinder Speed Range (rpm)	Straw Walkers	Cross-Shaker	Separation/Cleaning Area (sq. in.)	Grain Tank (bu)	Engine hp
Lanz										(DIN)
MD14Z/MD120	PTO	4'		15.5 x 31.5	1420	Rack	N	2381	---	---
MD180/MD195	PTO/ engine	6'/6'4"		19.75 x 55	910/1135		N		29	34G/34D
MD15S/MD18S	SP	6'/6'6"		15.5 x 31.5	545-1450	5	N	3565	16	30
MD150S	SP	6'		18 x 22.75	655-1380	3	N	2123/813	13	29G/32D
MD240S/MD250S MD25S	SP	8'		19.75 x 31.5	640-1270	4	N	4185/3223		50
MD260S/MD300S/ MD350S/MD35S/MD40S	SP	8.5'-10'		19.75 x 55	648-1250	4	N	4495	29	68
30 Series										(SAE)
330	SP	7'-8'3"	2	24 x 31	500-1100	3	N	3554/2220	50	52
360	PTO	7'-8'	2	24 x 31	500-1100	3	N	3554/2220	50	---
430	SP	8'-10'	2	24 x 31	500-1100	3	N	4373/3226	61	71
530	SP	10'-12'	3/4	24 x 41	500-1100	4	N	5832/4406	90	79
630	SP	10'-14'	2/3/4	24 x 41	500-1100	4	N	5863/4419	90	100
730	SP	12'-16'	4	24 x 51	500-1100	5	N	7322/5580	96	115
900 Series										
930	SP	8'-10'		24 x 31	500-1100	3	N	4386/2883	50	71
940	SP	8'6"-12'		24 x 31	500-1100	3	Y	5161/3487	61	79
950	SP	10'-14'		24 x 41	500-1100	4	Y	5843/5626	61	105
960/960 REV	SP	10'-16'		24 x 41	500-1100	4	Y	5843/5626	71	111
970/970 REV	SP	12'-18'		24 x 51	500-1100	5	Y	8602/6045	96	132
900 Series Updated										
925/932	SP	7'8"-8'5"'		24 x 31	500-1100	3	N	4386/2883	50	61
935	SP	8'5"-10'		24 x 31	500-1100	3	N	4386/2883	50	81
942	SP	8'5"-10'	3	24 x 31	470-1160	3	N	5161/2883	50	81
945	SP	8'6"-12'	3	24 x 31	470-1160	3	Y	5161/4107	61	93
952	SP	8'6"-14'	3/4	24 x 41	470-1160	4	N	5843/5626	77	93
955	SP	10'-16'	3/4	24 x 41	470-1160	4	Y	5843/5626	89	117
965/965H/968H	SP	10'-16'	4	24 x 41	470-1160	4	Y	5843/5626	112	1138
975/975 REV/975H4	SP	10'-18'	4/5/6	24 x 51	470-1160	5	Y	7316/7130	121	167
985/985 REV	SP	14'-18'	5/6	24 x 61	217-1200	6	Y	8773/8246	121	190
985Hydro4										217
1000 Series										
1032/1042	SP	8'5"-10'		24 x 41	470-1160	4	N	5735/3906	60	81
1051 (Australia)	PTO	22' & 25' comb, 25' open		24 x 41	470-1160	4	N	6882/5626	225	---
1052	SP	10'-14'	3/4	24 x 41	470-1160	4	N	6882/5626	86	93
1055	SP	10'-16'	4	24 x 41	470-1160	4	Y	6882/5626	86	117
1065	SP	10'-16'	4	24 x 41	470-1160	4	Y	6882/5626	123	138
1065A (Argentina)	SP	16'	4	24 x 41	380-1100	4	N	6882/5626	112	138
1068H	SP	12'-16'	4	24 x 41	470-1160	4	Y	6882/5626	112	138
1072	SP	12'-18'		24 x 51	565-1135	5	Y	8600/7130	137	138
1075/1075Hydro4	SP	12'-18'	4/5	24 x 51	225-1135	5	Y	8600/7130	137/171	167
1075A (Argentina)	SP	18'		24 x 51	380-1100	5	N	8600/7130	137	167
1085/1085Hydro4	SP	12'-18'	5/6	24 x 61	225-1135	6	Y	10323/8246	137/171	188/217
1100 & 1100 S II Series										
1133	SP	8'6"		24 x 41	340-1060	4	N	4743/3906	60	70
1144	SP	8'6"-10'		24 x 41	340-1060	4	N	4743/3906	60	81
1155	SP	10'-16'	4	24 x 41	470-1160	4	N	6048/4752	85	93
1157	SP	10'-16'	4	24 x 41	470-1160	4	Y	6048/4752	100	117
1158	SP	10'-16'		24 x 41	470-1160	4	Y	6048/4752	100	127
1166/1166H4/1169H	SP	10'-16'	4/5	24 x 41	190-1160	4	Y	6290/5626	125/112	138
1174 S II	SP	12'-18'		24 x 51	190-1160	5	Y	7817/6897	137	138 167
1177/1177Hydro4 S II	SP	12'-18'	5/6	24 x 51	158-1100	5	Y	7817/6897	137/171 160/171	167/205 182/216
1188/1188Hydro4 S II 1188Hydro4	SP	14'-20'	5/6	24 x 61	158-1100	6	Y	9435/7632	137/171 160/171	205/228 239

391

Additional Reading

JOHN DEERE

* *John Deere Tractors and Equipment*, Volume One, 1837-1959, Don Macmillan with Russell Jones

* *How Johnny Popper Replaced the Horse*, Ralph Hughes and Don Huber

* *The Toy and the Real McCoy*, Ralph C. Hughes

* *John Deere Tractors, 1918-1987*, Deere & Company

  John Deere's Company, Wayne Broehl

* *Johnny Tractor and His Pals*, for children, Louise Price Bell, Roy A. Bostrom

* *Corny Cornpicker Finds a Home*, for children, Lois Zortman Hobbs, Roy A. Bostrom

* *Family Reunion*, for children, Lois and J. R. Hobbs, Kris Carr, Roy A. Bostrom

* *John Deere Tractors: Big Green Machines in Review*, Henry Rasmussen

  Two-Cylinder (Bi-monthly), 531 Commercial, 7th Floor, Suite 700, Waterloo, IA 50701

  Green Magazine (Monthly), RR 1, Bee, NE 68314

OTHER COMPANIES

* *Ford Tractors: N Series, Fordson, Ford and Ferguson, 1914-1954*, Robert N. Pripps

* *Massey-Ferguson Tractors*, Michael Williams

* *Allis-Chalmers Agricultural Machinery*, Bill Huxley

* *150 Years of International Harvester*, Charles H. Wendel

* *Minneapolis-Moline Tractors 1870-1969*, Charles H. Wendel

* *150 Years of J. I. Case*, Charles H. Wendel

* *Case Tractors: Steam to Diesel*, Dave Arnold

* *Benjamin Holt and Caterpillar: Tracks and Combines*, Reynold M. Wik

INDUSTRY

* *The Agricultural Tractor: 1855-1950*, R.B. Gray

* *Farm Tractors: 1950-1975*, Les Larsen

* *The Grain Harvesters*, Graeme Quick and Wesley Buchele

* *Australian Tractors*, Graeme Quick

* *International Directory of Model Farm Tractors*, Raymond E. Crilley and Charles E. Burkholder

  Official Guide, Tractors and Farm Equipment, (Semi-annual), North American Equipment Dealers Association, 10877 Watson Rd., St. Louis, MO 63127-1081

  Implement and Tractor Red Book (Annual), P.O. Box 1420, Clarksdale, MS 38614

  Implement and Tractor (Bi-monthly), P.O. Box 1420, Clarksdale, MS 38614

  Farm Industry News (Bi-monthly), 7900 International Drive, Minneapolis, MN 55425

  Farm Show Magazine (Bi-monthly), Box 1029, 20080 Kenwood Trail, Lakeville, MN 55044

  Outdoor Power Equipment (Monthly), Chilton Company, 1 Chilton Way, Radnor, PA 19089

  Construction Equipment (Monthly), Cahners Publishing Co., 1350 Touhy Avenue, Box 5080, Des Plaines, IL 60017

  The Toy Farmer (Monthly), RR 2, Box 5, LaMoure, ND 58458

---

* Books can be ordered from the American Society of Agricultural Engineers (ASAE), Dept. 1555, 2950 Niles Road, St. Joseph, MI 49085-9659. Voice: (616) 429-0300; FAX: (616) 429-3852.
  If you wish to be kept current on equipment history publications, just ask for your free copy of the *Looking Back* catalog.